Chemical Ionization Mass Spectrometry

2nd Edition

Alex. G. Harrison
Department of Chemistry
University of Toronto
Toronto, Ontario
Canada

CRC Press

Boca Raton   Ann Arbor   London   Tokyo

Library of Congress Cataloging-in-Publication Data

Harrison, Alexander G.
     Chemical ionization mass spectrometry / Alex G. Harrison. — 2nd ed.
         p. cm.
     Includes bibilographical references and index.
     ISBN 0-8493-4254-6
     1. Chemical ionization mass spectrometry. I. Title.
QD96.M3H37   1992
543'.0873—dc20

                                  91-47678
                                  CIP

© 1992 by CRC Press, Inc.

International Standard Book Number 0-8493-4254-6

Library of Congress Card Number 91-47678
Printed in the United States 1 2 3 4 5 6 7 8 9 0
Printed on acid-free paper

PREFACE

The technique of ionizing a sample of molecules by gas-phase ion/molecule reactions was first reported in 1966 as an outgrowth of fundamental studies in gas-phase ion chemistry. In this pioneering paper, Munson and Field suggested that the method, which they called chemical ionization, might have useful analytical applications. The past 25 years have proven this prophecy to be abundantly correct. Chemical ionization provides information which is complementary to electron ionization and the two methods together are particularly powerful. In addition, is some cases, chemical ionization provides a more sensitive method for detecting trace levels of organic compounds. As a result, chemical ionization is a widely used ionization method in mass spectrometry. It also has found extensive applications in structural elucidation studies and quantitative analytical studies in many branches of chemistry and biochemistry and in medical and environmental areas. At the same time chemical ionization studies have provided a large body of information concerning gas-phase ion chemistry.

The literature of chemical ionizaton mass spectrometry is very large; over 1100 articles have been published using the phrase "chemical ionization" in the title since the first edition of this monograph was completed in early 1982. Thus, it appeared appropriate to update the book. The published papers range from basic studies of the ionization processes to applied analytical problems where chemical ionization was the method of choice. Clearly, it is impossible to review all this work. Rather, I have focused on the basic work which forms the body of knowledge upon which the applied studies are based. Since many of the scientists beginning work with chemical ionization do not have a background in gas-phase ion chemistry, I have, once more, provided an extensive review of the basic studies of ion/molecule reaction kinetics and thermodynamics, since these are the foundations upon which chemical ionization is based. At the same time I have not attempted to include the basic principles of mass spectrometry; these have received excellent coverage in other books.

My understanding of gas-phase ion chemistry and chemical ionization mass spectrometry has benefited greatly from the work of many associates over the past thirty years, and I am indebted to them for their contributions. As usual, Frank Safian has proven invaluable in the drafting of the diagrams. I am particularly indebted to my wife, Barbara, who not only has provided the usual unfailing support but also has entered all the tables into the word processor far more skillfully than I could have done.

A. G. H.
Toronto, October 1991

THE AUTHOR

Alex. G. Harrison is Professor of Chemistry at the University of Toronto. He received his Ph.D. in chemistry from McMaster University in 1956. After postdoctoral work at McMaster and the National Research Council of Canada, Ottawa, he joined the staff of the Department of Chemistry, University of Toronto, as a Lecturer in 1959, becoming Professor of Chemistry in 1967. He has spent sabbatical leaves in the Department of Molecular Sciences, University of Warwick (1975), Institut de Chimie Physique, Ecole Polytechnique Fédérale, Lausanne (1982) and the Department of Chemistry and Biochemistry, University of Colorado (1989).

Dr. Harrison is a Fellow of the Chemical Institute of Canada and a member of the American Chemical Society, the Canadian Society for Mass Spectrometry and the American Society for Mass Spectrometry. He has served on the Board of Directors of the latter society and as Chairman of the Physical Chemistry Division and Member of Council of the Chemical Institute of Canada. From 1988 to 1990 he served as the inaugural President of the Canadian Society for Mass Spectrometry. He is on the Editorial Advisory Boards of *Organic Mass Spectrometry*, *Mass Spectrometry Reviews*, *International Journal of Mass Spectrometry and Ion Processes* and the *Journal of the American Society for Mass Spectrometry* and is an Associate Editor of the *Canadian Journal of Applied Spectroscopy*.

Dr. Harrison was an Alfred P. Sloan Fellow 1962-64 and a Killam Research Fellow 1985-87. He has authored or co-authored more than 200 papers in the area of mass spectrometry and gas-phase ion chemistry. His research interests lie in the chemistry of gas-phase ions and in the analytical applications of mass spectrometry.

TABLE OF CONTENTS

Chapter 6

Chapter 1

INTRODUCTION

I. PREFACE

The technique of chemical ionization (CI), first introduced by Munson and Field[1] in 1966, is a direct outgrowth of fundamental studies of ion/molecule interactions; as such the technique is based on the knowledge developed from these fundamental studies and makes use of the instrumentation developed for such studies. Since its initial introduction, chemical ionization mass spectrometry (CIMS) has developed into a powerful and versatile tool for the identification and quantitation of organic molecules and, consequently, has found extensive application in many branches of chemistry and biochemistry and in medical and environmental fields.

In CIMS, ionization of the sample of interest is effected by gas phase ion/molecule reactions rather than by electron impact, photon impact, or field ionization/desorption; frequently CIMS provides information which is complementary to these techniques rather than supplementary. Particularly in structure elucidation, it is common to determine both the electron ionization (EI) and CI mass spectra. Although much of the earlier work in CIMS utilized positive ion/molecule reactions, there has been, over the past 15 years, an increased interest in negative ion/molecule reactions as an ionization method. This later development of negative ion CIMS occurred, in part, because EI studies at low source pressures showed, in general, a low sensitivity for negative ion production as well as a strong dependence of negative ion mass spectra on the ionizing electron energy. In addition, many commercial mass spectrometers were not equipped to operate in the negative ion mode. This latter problem has been rectified and most commercial mass spectrometers now operate readily in the negative ion mode which, under CI conditions, often shows ion signals at least as strong as those observed in the positive ion mode.

The essential reactions in CI are given in general form in Reactions 1 to 3. Ionization of the reagent gas (present in large excess), frequently by electron impact, usually is followed by ion/molecule reactions involving the primary ions and the reagent gas neutrals and produces the CI reagent ion or reagent ion array (Reaction 1). This reagent ion or reagent ion array represents the stable

$$e + R \rightarrow R^{\pm} \tag{1}$$

$$R^{\pm} + A \rightarrow A_1^{\pm} + N_1 \tag{2}$$

1

$$A_1^{\pm} \rightarrow A_2^{\pm} + N_2$$
$$\rightarrow A_3^{\pm} + N_3$$
$$\rightarrow A_i^{\pm} + N_i$$

(3)

end product(s) of the ion/molecule reactions involving the reagent gas. Collision of the reagent ion(s), $R^{\pm}$, with the additive (present at low concentrations, usually <1% of the reagent gas) produces an ion, $A_1^{\pm}$ characteristic of the additive (Reaction 2). This additive ion may fragment by one or more pathways, as in Reaction 3, or, infrequently, react with the reagent gas; the final array of ions $A_1^{\pm}$ to $A_i^{\pm}$ when mass analyzed in the usual manner, provides the CI mass spectrum of the additive gas A as effected by the reagent gas R.

A large part of the usefulness of CIMS rests with the fact that a large variety of reagent gases and, hence, reagent ions can be employed to effect ionization. To a considerable extent, the choice of reagent systems can be tailored to the problem to be solved. The problems amenable to solution by the CI technique can be divided into three rough categories: (1) molecular mass determination, (2) structure elucidation, and (3) identification and quantitation.

II. MOLECULAR MASS DETERMINATION

When a gaseous sample of a polyatomic species is bombarded with electrons at sample pressures of $\sim10^{-5}$ torr, the initial electron/molecule interaction produces an assembly of molecular ions with internal energies ranging from 0 to about 10 eV. The energized molecular ions fragment to form the fragment ions observed in the EI mass spectrum.[2] For some classes of compounds, the critical energy required for fragmentation of the molecular ion is extremely low (or zero) with the result that no molecular ions are seen in the mass spectrum; in effect, fragmentation is so facile that no molecular ions survive for the $\sim10^{-5}$ s that elapses before mass analysis occurs. When this happens, the molecular mass can be established only with considerable difficulty by EI methods. The aim of CI in these instances is to produce, by suitable choice of reagent ions, abundant ions characteristic of the sample and providing the molecular mass of the sample molecules.

In positive ion CI the most commonly used ionization reaction for such purposes has been proton transfer (Reaction 4) from the reagent ion(s), BH^+, to the sample molecule M, where ΔH_4 is given by the proton affinity (PA) of

$$BH^+ + M \rightarrow MH^+ + B, \Delta H_4$$

(4)

B less the PA of M. The extent of fragmentation of MH^+ depends on its internal energy which, in turn, depends on ΔH_4, the exothermicity of the proton transfer reaction. The magnitude of ΔH_4 and thus, the extent of fragmentation of MH^+

usually can be controlled by suitable choice of reagent gas; in addition, the stability of the even-electron MH^+ ion may be considerably greater than the stability of the odd-electron $M^{+\cdot}$ species formed in the EI process.

An analogous proton transfer (Reaction 5) can be utilized in negative ion CI; ΔH_5 is equal to $PA([M - H]^-) - PA(X^-)$

$$X^- + M \rightarrow [M - H]^- + HX \tag{5}$$

or, alternatively, is given by the gas phase acidity of M less the gas phase acidity of XH. Again, by suitable choice of reagent ion, the exothermicity of the proton transfer reaction can be made sufficiently small as to preclude extensive fragmentation of $[M - H]^-$. In addition, the exothermicity of the reaction appears to reside primarily in the bond formed between X and H and, therefore, is unavailable to promote fragmentation of $[M - H]^-$.

A number of other reactions also have been used to provide molecular mass information. These include hydride abstraction (Reaction 6), electron attachment (Reaction 7), electron transfer (Reaction 8), and adduct or cluster ion formation (Reaction 9). It should be noted that formation of cluster ions

$$X^+ + M \rightarrow [M - H]^+ + XH \tag{6}$$

$$e + M \rightarrow M^{-\cdot} \tag{7}$$

$$X^\pm + M \rightarrow M^\pm + X \tag{8}$$

$$X^\pm + M \rightarrow M \cdot X^\pm \tag{9}$$

usually requires third-body stabilization and the rates of such reactions may be considerably slower than the rates of bimolecular processes such as Reactions 4 to 8.

In the literature, ions characteristic of the molecular mass have been called quasimolecular ions independent of their exact identity as MH^+, $[M - H]^\pm$ or $M \cdot X^\pm$. This sloppy terminology, which has been criticized recently,[3] should be discontinued; the ions from which the molecular mass is established should be clearly identified and not hidden under a generic name.

III. STRUCTURE ELUCIDATION

When a sample of unknown chemical structure is being examined, one usually desires structural information as well as molecular mass information from the CI mass spectrum. Ideally, one would like to have an array of reagent ions which give characteristic reactions with each possible functional group, thus signalling the presence or absence of a particular group. Except for a very few cases, this ideal has not been attained and, consequently, other approaches must be used.

The majority of structural studies have used positive ion proton transfer CI methods. The major fragmentation reactions of the even-electron MH^+ ions formed in Reaction 4 involve elimination of stable neutral molecules, HY, where Y is a functional group such as OR, NR_2, halogen, etc., present in the molecule. This fragmentation mode frequently is quite different from the fragmentation modes of the odd-electron molecular ion formed by EI, and the structural information obtained by the two techniques often is complementary.

In those cases where the fragment ions formed by EI are structurally informative but no molecular mass information is provided by the EI mass spectrum, mixed proton transfer/charge exchange reagent systems can be used. The proton transfer reagent forms MH^+ ions indicative of molecular mass, while the charge exchange reagent, by dissociative charge exchange, gives the same fragment ions as are produced in EI since the initial ion/molecule interaction produces the odd-electron molecular ion.

The methodology of structure elucidation by negative ion CI is less developed, although general rules for the fragmentation of negative ions are being developed.[4] Increasingly, structural information is being derived by collision-induced dissociation, in a tandem mass spectrometer, of the MH^+ or $[M - H]^-$ ions formed in a gentle CI reaction.[5,6]

IV. IDENTIFICATION AND QUANTITATION

This application of mass spectrometry is undoubtedly the most common, whether employing EI or CI. The aim is to identify a known substance and to measure quantitatively the amount of this substance present in a frequently complex matrix. Although, in principle, it should be possible to selectively ionize the component of interest in the matrix by a suitable CI reaction, this approach has seen relatively little use. In the majority of studies, separation of the mixture in time is achieved by application of a chromatographic technique with direct introduction of the chromatographic effluent into the mass spectrometer ion source. The ionization reaction requires only that a relatively small number of ions characteristic of the component be produced in high yield to achieve maximum sensitivity of detection. In general, formation of two or more ions is preferred, since identification is made more certain by the requirements that the appropriate ion signals must show the same intensity/time profile as well as the appropriate relative intensities. Cairns et al.[7] recently have discussed in detail the criteria for confirmation of the presence of a component in a complex matrix.

An alternative to separation of the components in time is separation of the components by mass in so-called MS/MS experiments.[5,6] In this approach the complex mixture is ionized, usually by gentle methods, to produce characteristic ions such as MH^+ or $[M - H]^-$ from each component. Mass selection of the ion suspected to arise from the component of interest and collision-induced dissociation of this ion in a tandem mass spectrometer leads to a fragment ion or ions which are characteristic of the component. In practice, it has been found that some separation by chromatographic methods usually is advantageous when this approach to identification is used.

The problems of quantitation, particularly in gas chromatography work, have been discussed in detail[8] and will not be repeated here. Because of potential losses of the analyte during the workup of the sample, an internal standard of some type is added to the mixture at the earliest possible stage of sample handling. This internal standard should have chemical and physical properties similar to the component of interest so that the losses of the two will be the same. Knowing the amount of standard added, the amount of unknown present can be deduced from a comparison of the mass spectral response to the unknown and the standard.

V. SCOPE OF THE PRESENT WORK

These introductory comments serve to indicate the scope of the present monograph. CIMS rests firmly on the foundations established from fundamental studies of gas-phase ion chemistry. These foundations are discussed in Chapter 2. The techniques of mass spectrometry peculiar to CI studies are discussed in Chapter 3, although no attempt is made to review the basic principles of mass spectrometry; for these the reader is referred to other sources.[9–12] A particularly elegant comparison of various mass analyzers has appeared recently.[13] The reagent gases used in CIMS have been many, with various areas of utility; the more important are discussed in Chapter 4. Similarly, the range of compounds studied by chemical ionization mass spectrometry is vast. Those for which systematic studies have been made, allowing generalizations to be drawn, are discussed in Chapter 5. Finally, a number of more specialized topics are discussed in Chapter 6.

REFERENCES

1. **Munson, M.S.B. and Field, F.H.**, Chemical ionization mass spectrometry. I. General introduction, *J. Am. Chem. Soc.*, 88, 2621, 1966.
2. **Harrison, A.G. and Tsang, C.W.**, The origin of mass spectra, in *Biochemical Applications of Mass Spectrometry*, Waller, G.R., Ed., John Wiley & Sons, New York, 1972.
3. **Bursey, M.M.**, Comment to readers. Style and the lack of it, *Mass Spectrom. Rev.*, 10, 1, 1991.
4. **Bowie, J.H.**, The fragmentation of even-electron organic ions, *Mass Spectrom. Rev.*, 9, 349, 1990.

5. **McLafferty, F.W.**, Ed., *Tandem Mass Spectrometry*, John Wiley & Sons, New York, 1983.

6. **Busch, K.L., Glish, G.L. and McLuckey, S.A.**, *Mass Spectrometry/Mass Spectrometry: Techniques and Applications of Tandem Mass Spectrometry*, VCH Publishers, New York, 1988.

7. **Cairns, T., Siegmund, E.G., and Stamp, J.J.**, Evolving criteria for the confirmation of trace level residues in food and drugs by mass spectrometry. Part I, *Mass Spectrom. Rev.*, 8, 93, 1989. Part II, *Mass Spectrom. Rev.*, 8, 127, 1989.

8. **Millard, B.J.**, *Quantitative Mass Spectrometry*, Heyden and Son, London, 1978.

9. **Roboz, J.**, *Introduction to Mass Spectrometry*, Interscience, New York, 1968.

10. **Watson, J.T.**, *Introduction to Mass Spectrometry, 2nd. Edition*, Raven Press, New York, 1985.

11. **Chapman, J.R.**, *Practical Organic Mass Spectrometry*, John Wiley & Sons, Chichester, 1985.

12. **White, F.A. and Wood, G.M.**, *Mass Spectrometry: Applications to Science and Engineering*, John Wiley & Sons, New York, 1986.

13. **Brunnée, C.**, The ideal mass analyzer: fact or fiction, *Int. J. Mass Spectrom. Ion Processes*, 76, 125, 1987.

Chapter 2

FUNDAMENTALS OF GAS-PHASE ION CHEMISTRY

I. INTRODUCTION

From the earliest days of mass spectrometry, ions were observed which generally were agreed to have arisen from reactions between ions and neutral molecules. An ion of m/z 3 was observed by Dempster[1] in 1916 and identified as H_3^+ ; its formation by Reaction 1 was well established by 1925.[2,3] In 1928,

$$H_2^{+\cdot} + H_2 \rightarrow H_3^+ + H^\cdot \tag{1}$$

Hogness and Harkness[4] reported the formation of I_3^+ and I_3^- in iodine vapor subjected to electron impact. Ion/molecule reactions were observed in several other systems during the 1920s; the early work has been reviewed by Smyth[5] and by Thompson.[6] With improvements in instrumentation and techniques, particularly vacuum technology, the nuisance of secondary processes was largely eliminated, and studies of ion/molecule reactions largely ceased. The main interests in mass spectrometry during the period 1930 to 1950 lay in the physics of ionization and dissociation, in the determination of isotopic masses and abundances, and in the development of analytical mass spectrometry using primarily electron ionization.

The modern era of ion/molecule reaction studies began in the early 1950s when the ion CH_5^+, formed by the reaction

$$CH_4^{+\cdot} + CH_4 \rightarrow CH_5^+ + CH_3^\cdot \tag{2}$$

was discovered by Tal'roze and Lyubimova[7] in the Soviet Union and, independently soon thereafter, by Stevenson and Schissler[8] and Field, Franklin, and Lampe[9] in the U.S. The observation of CH_5^+ as a stable species aroused the interest of chemists concerned with structure and bonding, while the observation that Reaction 2 was considerably faster than reactions involving only neutral species suggested that ion/molecule reactions might play an important role in radiation chemistry and aroused the interest of radiation chemists. In addition, much improved equipment was available for the controlled study of ionic collision processes. As a result a number of studies of gas-phase ion/molecule reactions were undertaken. Since that beginning, the study of the products, distribution, rates, and equilibria of gas-phase ionic reactions has become a major field of scientific activity with applications in many diverse fields. In the course of these studies, many advances in instrumentation have been made. The advances in instrumentation and in the understanding of gas-phase ion chemistry have been reviewed in numerous articles[10–18] and books.[19–29]

The instrumentation developed for the study of ion/molecule reactions has led to the instrumentation for CIMS; this aspect is discussed in Chapter 3. The large body of kinetic and thermochemical data derived from the fundamental

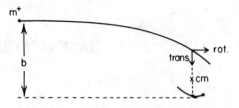

FIGURE 1.   Schematic ion/molecule collision.

studies constitutes the foundation upon which the chemistry of the chemical ionization technique is based; this body of data is reviewed in the remainder of this chapter.

II. ION/MOLECULE COLLISION RATES

The usefulness of an ion/molecule reaction in a chemical ionization system depends, in part, on the identity of the reaction products and, in part, on the rate of the reaction. The latter is important since only those reactions which are rapid can be expected to show adequate product yields. The upper limit to the reaction rate is given by the collision rate. The present section reviews the estimation from theory of collision rates of ions with nonpolar and with polar molecules.

A. Langevin Ion-Induced Dipole Theory

When an ion interacts with a nonpolar molecule it induces a dipole in the neutral; at moderately long range the only interaction of importance is the resultant ion-induced dipole interaction. Eyring et al.[30] first calculated in 1936 the classical capture collision cross section for the reaction $H_2^+ + H_2$ using absolute rate theory and a model of a structureless point charge interacting with a point-polarizable molecule. A general form of the capture collision cross section was derived by Vogt and Wannier,[31] and elaborated by Gioumousis and Stevenson[32] based on a model first developed by Langevin.[33] The theoretical development has been discussed extensively by a number of authors;[16,34–37] the present summary is based largely on the treatment of Su and Bowers.[36]

The theory calculates the collision cross section for an ion/molecule pair with a given relative velocity where both partners are assumed to be point particles with no internal energy and the interaction between the two is assumed to arise from ion-induced dipole forces only. Thus, for an ion and a molecule approaching each other with a relative velocity v and impact parameter b (Figure 1) the classical potential at an ion/molecule separation r is given by

$$V(r) = -\alpha q^2 / 2r^4 \qquad (3)$$

where α is the polarizability of the neutral and q is the charge on the ion. For $r < \infty$, the relative energy of the system, E_r, is the sum of the instantaneous kinetic energy and the potential energy:

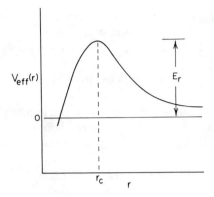

FIGURE 2. V_{eff} vs. r for critical impact parameter b_c.

$$E_r = \tfrac{1}{2}\mu v^2 = E_{kin}(r) + V(r) \tag{4}$$

where μ is the reduced mass. There are two components to $E_{kin}(r)$, the translational energy along the line of center of the collision, $E_{trans}(r)$, and the energy of relative rotation of the particles, $E_{rot}(r)$, where the latter is given by

$$E_{rot}(r) = L^2/2\mu r^2 = \mu v^2 b^2/2r^2 \tag{5}$$

where L is the classical orbital angular momentum of the two particles. The rotational energy is associated with a outwardly acting centrifugal force and the effective potential of the ion/molecule system can be represented as the sum of the central potential energy and this centrifugal potential energy

$$V_{eff}(r) = -\left(\alpha q^2/2r^4\right) + \left(L^2/2\mu r^2\right) \tag{6}$$

The total relative energy of the system, thus, is

$$E_r = E_{trans}(r) + V_{eff}(r) \tag{7}$$

The variation of $V_{eff}(r)$ with r at constant E_r depends on the value of the impact parameter b. For b = 0, there is no contribution from the centrifugal potential and $V_{eff}(r)$ is attractive for all values of r. When b > 0 the centrifugal potential creates a so-called "centrifugal barrier" to a capture collision. The special case (b = b_c) where the centrifugal barrier height equals E_r is shown in Figure 2; at r_c, $V_{eff}(r) = E_r$ and thus, from Equation 7 $E_{trans}(r) = 0$ and the particles will orbit the scattering center with a constant ion/molecule separation r_c. For all impact parameters less than b_c, a capture collision will occur, where a capture collision is defined as one in which the particles have appropriate energy and appropriate impact parameters to pass through r = 0. (For real ions and molecules there would, of course, be some finite minimum value of r > 0).

For impact parameters greater than b_c the centrifugal barrier prevents capture and the particles are simply scattered at large values of r.

The capture collision cross section at a given velocity v is defined by

$$\sigma(v) = \pi b_c^2 \tag{8}$$

Thus, the critical impact parameter b_c is the impact parameter such that $V_{eff}(r) = E_r$ and can be evaluated from the conditions that $\partial V_{eff}(r)/\partial r = 0$ and $V_{eff}(r) = E_r$ at $r = r_c$, i.e.,

$$\partial V_{eff}/\partial r = 0 = \left(L^2/\mu r^3\right) + \left(2\alpha q^2/r^5\right) \tag{9}$$

$$V_{eff}(r) = \left(L^2/2\mu r^2\right) - \left(\alpha q^2/2r^4\right) = E_r = \tfrac{1}{2}\mu v^2 \tag{10}$$

These restrictions lead to

$$r_c = \left(2q/b_c v\right)\left(\alpha/\mu\right)^{\frac{1}{2}} \tag{11}$$

$$b_c = \left(4\alpha q^2/\mu v^2\right)^{\frac{1}{4}} \tag{12}$$

which leads to the distance of closest approach $r_c = b_c/2^{1/2}$ and to a capture collision cross section and collision rate constant given by

$$\sigma_c(v) = \pi b_c^2 = \left(2\pi q/v\right)\left(\alpha/\mu\right)^{\frac{1}{2}} \tag{13}$$

$$k_c = v\,\sigma_c(v) = 2\pi q\left(\alpha/\mu\right)^{\frac{1}{2}} \tag{14}$$

Hence, if only ion-induced dipole interactions are involved, classical collision theory predicts that the capture cross section should vary inversely as the relative velocity of the colliding pair, while the capture rate constant, frequently called the Langevin rate constant, should be independent of the relative velocity and the temperature. As will be seen in the following, Equation 14 predicts reasonably well the maximum rate constants for ion/molecule reactions involving nonpolar molecules. This indicates that reaction occurs on every collision for many ion/molecule pairs; consequently, there can be no activation energy for the reaction. The rate constant predicted from Equation 14 is of the order of 1×10^{-9} cm^3 molecule^{-1} s^{-1} or about 6×10^{11} L mol^{-1} s^{-1}.

B. Average Dipole Orientation (ADO) Theory

Although Equation 14 predicts reasonably well the maximum rate constants for reactions involving nonpolar molecules, it seriously underestimates the rate constants of most ion/molecule reactions involving polar molecules. For these cases, it was shown first by Moran and Hamill[38] that ion dipole forces could not be ignored. Where such forces are important, the effective potential, analogous to Equation 6, becomes

$$V_{eff}(r) = \left(L^2/2\mu r^2\right) - \left(\alpha q^2/2r^4\right) - \left(\mu_D \cos\theta/r^2\right) \tag{15}$$

where μ_D is the dipole moment and θ is the angle the dipole makes with the center of collision. Hamill and colleagues[38,39] made the simplifying assumption that the dipole "locks in" on the ion ($\theta = 0$) and derived, by the approach outlined above, the capture collision rate constant:

$$k_{LD}(v) = \left(2\pi q/\mu^{\frac{1}{2}}\right)\left[\alpha^{\frac{1}{2}} + \left(\mu_D/v\right)\right] \tag{16}$$

where the subscript LD refers to the locked-in dipole approximation. In this case, the rate constant depends on the relative velocity of the colliding pair and gives[40] for a Maxwell-Boltzmann distribution of relative velocities:

$$k_{LD}(\text{thermal}) = \left(2\pi q/\mu^{\frac{1}{2}}\right)\left[\alpha^{\frac{1}{2}} + \mu_D\left(2/\pi k_B T\right)^{\frac{1}{2}}\right] \tag{17}$$

where k_B is Boltzmann's constant and T is the absolute temperature.

It has been shown[40-43] that Equations 16 and 17 seriously overestimate the ion-dipole effect on ion/molecule reaction rate constants; presumably locking-in of the dipole does not occur. Thus, a more reasonable expression for the thermal energy rate constant is

$$k_{ADO}(\text{therm}) = \left(2\pi q/\mu^{\frac{1}{2}}\right)\left[\alpha^{\frac{1}{2}} + C\mu_D\left(2/\pi k_B T\right)^{\frac{1}{2}}\right] \tag{18}$$

where C, the dipole locking constant, reflects the extent to which the dipole of the molecule orients itself with the incoming charge and may have values from 0 (no alignment) to 1 (locking-in). Bowers and colleagues have developed the average dipole orientation (ADO) theory of ion/polar molecule collisions in which they calculate by classical statistical theory either the average orientation angle θ[42,43] or the average cos θ (i.e., the average effective potential of the ion/polar molecule system).[44] The two approaches give similar results. From the results of such calculations, they deduced that the dipole locking constant C of Equation 18 can be parameterized. It turns out that, at constant temperature, C is a function of $\mu_D/\alpha^{1/2}$ only. Values of C as a function of $\mu_D/\alpha^{1/2}$, covering a temperature range of 150 to 500 K, have been presented;[45] typical plots are shown in Figure 3. For large values of $\mu_D/\alpha^{1/2}$, C reaches a limiting value of ~0.26. Thus, the effect of ion dipole interactions is much less than predicted from the simple model involving locking-in of the dipole (C = 1), although enhancement of the rate constant by a factor of 2 to 4 over the ion-induced dipole value is possible for reactions where the molecule has a large dipole moment.

A number of studies,[42,43,46,47] principally involving simple proton transfer reactions have shown that the ADO theory in the form of Equation 18, with the

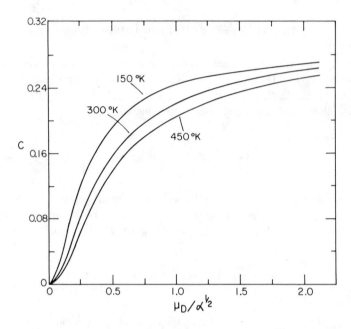

FIGURE 3. Dipole locking constant C as a function of $\mu_D/\alpha^{1/2}$ and temperature. Data from Su and Bowers.[45]

value of C derived from $\mu_D/\alpha^{1/2}$, adequately predicts the maximum rate constants for ion/polar molecule reactions. These will be discussed in more detail in the following sections. The ADO theory also predicts a much smaller dependence of rate constant on temperature than does the locked dipole expression, Equation 17; this also has been confirmed experimentally.[48,49] Further extensions of the theory to include conservation of angular momentum (AADO theory)[50] and to include ion-quadrupole interactions (AQO theory)[51] have been reported. These normally are relatively small effects and the reader is referred to the original publications or to the review by Su and Bowers[36] for details.

III. POSITIVE ION/MOLECULE REACTIONS

A wide variety of ion/molecule reactions involving positive ions have been studied. The present review will concentrate primarily on the type of product ions formed and on the reaction rates. The classification system used, while essentially arbitrary, is based largely on the identity of the product ions formed. The rate constants listed in this review are illustrative of the results available; a complete tabulation of kinetic data to 1986 is available.[52]

A. Charge Exchange
The interaction of a positive ion with a neutral molecule may lead to charge exchange, Reaction 19; the reaction will be exothermic if the

TABLE 1
Recombination Energies of Gaseous Ions[a]

Ion	RE (eV)[a]
$Ne^{+}\,^2P_{3/2}$	21.6
$^2P_{1/2}$	21.7
$Ar^{+}\,^2P_{3/2}$	15.8
$^2P_{1/2}$	15.9
N_2^{+}	15.3
$Kr^{+}\,^2P_{3/2}$	14.0
$^2P_{1/2}$	14.7
$CO^{+\cdot}$	14.0
$CO_2^{+\cdot}$	13.8
$Xe^{+}\,^2P_{3/2}$	12.1
$^2P_{1/2}$	13.4
$COS^{+\cdot}$	11.2
$CS_2^{+\cdot}$	~9.5 or 10.0
NO^{+}	8.5 or 9.5
$C_6H_6^{+\cdot}$	9.3

[a]   From Lindholm.[53,54]

$$X^{+\cdot} + M \rightarrow M^{+\cdot} + X \qquad \Delta H = IE(M) - RE\left(X^{+\cdot}\right) \qquad (19)$$

recombination energy (RE) of the reactant ion is greater than the ionization energy (IE) of the neutral M. The RE of $X^{+\cdot}$ is defined as the exothermicity of the gas phase reaction

$$X^{+\cdot} + e \rightarrow X \qquad -\Delta H = RE\left(X^{+\cdot}\right) \qquad (20)$$

For monatomic ions, the RE has the same numerical value as the ionization of the neutral; this is not necessarily so for diatomic and polyatomic species since the product neutral in these cases may have excess vibrational/rotational energy or, in some cases, may be formed in an excited electronic state. A selection of RE, taken from the summaries by Lindholm,[53,54] is presented in Table 1. For polyatomic M, the exothermicity of the charge exchange reaction appears to remain largely as internal energy of the $M^{+\cdot}$ product.[55] If this internal energy is sufficiently large the $M^{+\cdot}$ ion may undergo fragmentation; the fragmentation reactions observed will be the same as those observed following electron ionization since, in both cases, fragmentation commences from the odd-electron molecular ion. The important difference is that the molecular ions formed by electron ionization will have a distribution or range of internal energies while those formed by charge exchange will have an internal energy determined by the exothermicity of Reaction 19. Since the relative importance of the different fragmentation channels is dependent on the internal energy, the product distribution from charge exchange reactions will depend on the exothermicity of the reaction (see Section II.C, Chapter 4).

TABLE 2
Rate Constants for Charge Exchange Between Rare Gas Ions and Nonpolar Molecules

Reactants	k_{react} (cm^3 molecule^{-1} S^{-1} $\times$ 10^9)	k_{coll} (cm^3 molecule^{-1} S^{-1} $\times$ 10^9)
Ne$^+$ + CH$_4$	0.07[a]	1.26
Kr$^+$ + CH$_4$	1.26[a]	1.03
	1.03[b]	—
	1.20[c]	—
Ar$^+$ + CH$_4$	1.34[a]	1.12
	0.98[b]	—
	1.10[c]	—
Ne$^+$ + C$_2$H$_6$	0.98[a]	1.40
Ar$^+$ + C$_2$H$_6$	1.07[a]	1.16
Kr$^+$ + C$_2$H$_6$	0.75[a]	1.03
Xe$^+$ + C$_2$H$_6$	0.84[a]	0.98
Ne$^+$ + C$_3$H$_8$	1.21[a]	1.55
Ar$^+$ + C$_3$H$_8$	1.20[a]	1.25

[a]   From Bowers and Elleman.[56]
[b]   From Ausloos et al.[57]
[c]   From Li and Harrison.[58]

Charge exchange reactions of polyatomic molecules have been studied extensively using tandem mass spectrometers.[53,54] However, such studies have involved reactant ions with >3 eV kinetic energy and have been more concerned with product distributions than with the measurement of reaction rate constants. More recently, charge exchange reactions at nearer thermal energies have been studied using chemical ionization techniques,[55] but again the interest has been in product ion distributions rather than reaction rates. Consequently, the information concerning the rate constants of charge exchange reactions at low kinetic energies is rather limited. Table 2 summarizes rate constants measured for charge exchange involving rare gas ions and some simple alkanes. The table also compares the measured reaction rate constants with collision rate constants calculated by Equation 14. For no case does the measured rate constant exceed the collision rate constant. With the exception of the reaction Ne$^+$ + CH$_4$ the efficiencies of the reactions are quite high.

Table 3 records rate constants measured for charge exchange between the ions Ne$^+$, Ar$^+$, N$_2^+$, Kr$^+$ and CO$^+$ and the polar molecules dimethyl ether (μ_D = 1.30D), methyl alcohol (μ_D = 1.71D), and acetone (μ_D = 2.88D). Also included in the table are the collision rate constants calculated from the ADO theory, Equation 18. The latter have been separated into the ion-induced dipole and ion-permanent dipole contributions to illustrate the magnitude of each. The ion-permanent dipole contribution varies with the dipole moment of the neutral, and for the most polar molecule, acetone, is equal in magnitude to the ion-induced dipole contribution. Again, within experimental error, the measured reaction rate constants do not exceed the collision rate constants, although in

TABLE 3
Rate Constants for Charge Exchange Reactions Involving Polar Molecules

Reactants	$k_{calc'd}$ (cm^3 molecule^{-1} s^{-1} × 10^9)			$k_{expt'l}$[a] (cm^3 molecule^{-1} s^{-1} × 10^9)
	Ion-induced dipole	Ion-dipole	Total	
Ne$^+$ + (CH$_3$)$_2$O	1.44	0.51	1.95	1.66
Ne$^+$ + CH$_3$OH	1.21	0.93	2.14	1.36
Ne$^+$ + (CH$_3$)$_2$CO	1.54	1.52	3.06	3.15
Ar$^+$ + (CH$_3$)$_2$O	1.16	0.42	1.58	1.48
Ar$^+$ + CH$_3$OH	1.01	0.77	1.78	1.55
Ar$^+$ + (CH$_3$)$_2$CO	1.22	1.21	2.43	2.56
N$_2^+$ + (CH$_3$)$_2$O	1.29	0.46	1.75	1.95
N$_2^+$ + CH$_3$OH	1.10	0.84	1.94	1.75
N$_2^+$ + (CH$_3$)$_2$CO	1.37	1.35	2.72	2.76
Kr$^+$ + (CH$_3$)$_2$O	0.99	0.35	1.34	1.43
Kr$^+$ + CH$_3$OH	0.88	0.68	1.56	1.72
Kr$^+$ + (CH$_3$)$_2$CO	1.02	1.00	2.02	1.79
CO$^+$ + (CH$_3$)$_2$O	1.29	0.46	1.75	1.94
CO$^+$ + CH$_3$OH	1.10	0.84	1.94	1.83
CO$^+$ + (CH$_3$)$_2$CO	1.37	1.35	2.72	2.83

[a] Uncertainty ±0.2, data from Li and Harrison.[59]

most cases the reaction efficiency is high. The latter point is emphasized because, in principle, the electron transfer involved in Reaction 19 could occur over larger distances than are implied in capture or orbiting collisions; as a consequence, the observed rate constants could be larger than the calculated capture collision rate constants. There appears to be little concrete evidence for such an electron jump mechanism except possibly for resonant charge transfer reactions in the rare gases of the type A$^+$ + A. Thus, it appears that for polyatomic species, the rate constants for most exothermic charge exchange reactions will be close to the calculated capture collision limit.

B. Proton and Hydrogen Atom Transfer
The self-protonation reaction

$$M^{+\cdot} + M \rightarrow MH^+ + [M - H]^{\cdot} \tag{21}$$

is an ubiquitous reaction for many classes of compounds. Indeed, it is so prevalent that in many cases MH$^+$ ions may be observed under electron ionization conditions unless precautions are taken. (Fragment ions produced by electron ionization also may react with M to produce MH$^+$). Reaction 21 may occur either by transfer of H$^+$ from M$^{+\cdot}$ to M or by transfer of an H atom from M to M$^{+\cdot}$; the two reactions cannot be distinguished in a conventional single-source experiment. Using a tandem mass spectrometer, Abramson and Futrell[60]

TABLE 4
Reaction Rate Constants for Self-Protonation Reactions
$$M^{+\cdot} + M \rightarrow MH^+ + [M-H]^-$$

$M^{+\cdot} + M$	k_{react} $(cm^3 \ molecule^{-1} \ s^{-1} \times 10^9)$	k_{coll} $(cm^3 \ molecule^{-1} \ s^{-1} \times 10^9)$
$CH_4^{+\cdot} + CH_4$	1.11[a]	1.35
$NH_3^{+\cdot} + NH_3$	1.5–2.2[a]	2.10
$H_2O^{+\cdot} + H_2O$	2.05[a]	2.29
$CH_3OH^{+\cdot} + CH_3OH$	2.53[b]	1.78
$CH_3NH_2^{+\cdot} + CH_3NH_2$	1.22[c]	1.85
$CH_3F^{+\cdot} + CH_3F$	1.87[d]	1.89
$CH_3Cl^{+\cdot} + CH_3Cl$	1.53[d]	1.63
$CH_3SH^{+\cdot} + CH_3SH$	0.77[e]	1.43

[a] Data from Huntress and Pinnizotto.[61]
[b] Data from Gupta et al.[40]
[c] Data from Hellner and Sieck.[62]
[d] Data from Herod et al.[63]
[e] Data from Solka and Harrison.[64]

studied the reaction between $CD_4^{+\cdot}$ and CH_4. They found that D^+ transfer was favored over H atom transfer by approximately a factor of two. Similar results, showing that H^+ transfer is favored over H atom transfer, have been obtained by Huntress and Pinnizotto[61] for the reaction pairs $H_2O^{+\cdot} + H_2O$ and $NH_3^{+\cdot} + NH_3$ using isotopic labelling and ion cyclotron resonance (ICR) techniques. A further interesting observation from the $CD_4^+ + CH_4$ experiment is that only CD_4H^+ and CH_4D^+ product ions are observed with no isotopically mixed species such as $CH_3D_2^+$. This result indicates that, if a collision complex is formed, it is not sufficiently long-lived for isotopic interchange between the reaction partners to occur. This appears to be generally true for H^+/H atom transfer reactions and contrasts with the results obtained for condensation reactions which are discussed in Section III.D.

Table 4 records the measured reaction rate constants for a number of self-protonation reactions and compares these with the calculated capture collision rate constants. With the exception of the first entry, these were calculated from the ADO theory using Equation 18. The rate constant for the reaction between NH_3^+ and NH_3 is known to depend on the vibrational state of the reactant ion.[61] In general, reaction occurs on 50 to 90% of collisions for this type of reaction. These reactions are important in the preparation of gaseous Brønsted acid reagent ions.

Of even more interest in the present context are proton transfer reactions from gaseous Brønsted acids to neutral molecules, Reaction 22; this is a

$$BH^+ + M \rightarrow MH^+ + B \qquad (22)$$

common ionization reaction in CIMS. Table 5 records experimentally mea-

TABLE 5
H⁺/D⁺ Transfer from Gaseous Brønsted Acids

Reactants	k_{react} (cm³ molecule⁻¹ s⁻¹ × 10⁹)	k_{coll} (cm³ molecule⁻¹ s⁻¹ × 10⁹)
$CD_5^+ + C_2H_6$	1.07[a]	1.39
$CH_5^+ + C_3H_8$	1.54[b]	1.68
$CH_5^+ + C_2H_4$	1.51[c]	1.49
$CH_5^+ + CH_3NH_2$	2.51[d]	—
	2.25[e]	1.98
$CH_5^+ + (CH_3)_2NH$	2.25[d]	—
	2.15[e]	1.93
$CD_5^+ + CH_3Cl$	2.95[d]	2.27
$CH_5^+ + C_2H_5Cl$	3.02[f]	2.66
$CH_5^+ + n\text{-}C_5H_{11}Cl$	3.29[f]	3.03
$CD_5^+ + CH_3OH$	1.66[d]	1.98
$C_2H_5^+ + CH_3NH_2$	1.82[d]	—
	1.87[e]	1.70
$C_2H_5^+ + (CH_3)_2NH$	1.83[d]	—
	1.88[e]	1.62
$C_2D_5^+ + CH_3SH$	1.96[g]	1.74
$C_3H_7^+ + CH_3NH_2$	1.65[e]	1.63
$C_3H_7^+ + (CH_3)_2NH$	1.64[e]	1.40
$C_4H_9^+ + CH_3NH_2$	1.43[h]	1.54
$C_4H_9^+ + (CH_3)_2NH$	1.38[h]	1.40

[a] From Harrison, Heslin and Blair.[65]
[b] From Solka, Lau and Harrison.[66]
[c] From Solka and Harrison.[67]
[d] From Harrison, Lin and Tsang.[47]
[e] From Su and Bowers.[43]
[f] From Su and Bowers.[68]
[g] From Solka and Harrison.[64]
[h] From Su and Bowers.[69]

sured rate constants for proton transfer from CH_5^+, $C_2H_5^+$, $C_3H_7^+$ and $C_4H_9^+$ to a variety of small molecules and compares these rate constants with the calculated capture collision rate constants. The first three entries represent proton transfer to nonpolar molecules with the collision rate constants calculated from the polarization theory by Equation 14; the remaining entries represent proton transfer to polar molecules with the collision rate constants calculated from the ADO theory using the appropriate values of C in Equation 18.

These exothermic proton transfer reactions are all highly efficient with reaction occurring on essentially every collision. Indeed, the ADO theory appears to slightly (~20%) underestimate the collision rate. Better agreement is obtained using a combined variational transition state theory/classical trajectory approach to the calculation of the collision rate constants.[70,71] The high efficiency of exothermic proton transfer reactions also has been established in other studies using a variety of gaseous Brønsted acids.[46,72,73] Although all of these studies have involved rather simple neutral reactants, there is no reason

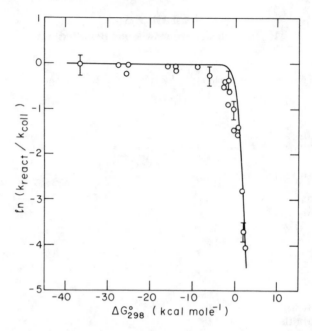

FIGURE 4.   Efficiency of proton transfer as a function of ΔG^0_{react}. (From Bohme, D.K., Mackay, G.I., and Schiff, H.I., *J. Chem. Phys.*, 73, 4976, 1980. With permission.)

to believe that exothermic proton transfer to complex molecules will not be equally efficient. For such complex molecules, with high polarizabilities and, frequently, dipole moments, proton transfer rate constants in the range 2 to 4×10^{-9} cm^3 molecule^{-1} s^{-1} can be expected.

As the proton transfer reaction becomes thermoneutral, the reaction efficiency drops[74,75] and becomes very low for endothermic or endoergic reactions. This is illustrated by the plot in Figure 4 of the reaction efficiency (k_{react}/k_{coll}) as a function of ΔG^0_{react} for a variety of proton transfer reactions as reported by Bohme et al.[75] The moral is that for maximum ionization efficiency in proton transfer CI one requires an exothermic protonation reaction. However, the more exothermic the protonation reaction the more extensive will be the fragmentation of MH$^+$. This aspect is discussed in detail in Chapter 4.

C. Negative Ion Transfer Reactions

The hydride ion transfer reaction

$$X^+ + M \rightarrow [M - H]^+ + XH \tag{23}$$

was first discovered by Field and Lampe[76] in a study of the ion/molecule reactions occurring on electron ionization of the C_2 to C_6 alkanes; in these systems, the reactant ions were the lower alkyl ions produced by dissociative ionization of the alkane. Such reactions lead, for example, to the formation of the $C_4H_9^+$ reactant ion as the major ionic species in the high pressure mass

TABLE 6
Hydride Ion Abstraction Reactions of Alkanes
$$X^+ + M \rightarrow XH + [M-H]^+$$

M/X=	$C_2H_5^+$	$i\text{-}C_3H_7^+$	$t\text{-}C_4H_9^+$	CF_3^+	NO^+
C_3H_8	6.3	1.2	—	5.9	—
$n\text{-}C_4H_{10}$	8.4	5.6	—	7.5	<0.02
$i\text{-}C_4H_{10}$	10	4.2	—	5.5	4.6
$n\text{-}C_5H_{12}$	10.9	8.3	<0.004	8.9	<0.05
$i\text{-}C_5H_{12}$	10.9	7.5	0.25	8.0	7.8
$c\text{-}C_6H_{12}$	16	11	<0.0005	—	1.3
$2\text{-}CH_3C_5H_{11}$	—	—	0.39	—	11.8
$3\text{-}CH_3C_5H_{11}$	—	—	0.27	—	12.1
$n\text{-}C_7H_{16}$	—	—	—	11.8	0.47
$3\text{-}CH_3C_6H_{13}$	—	—	0.45	—	14.8

Note: k (cm^3 molecule^{-1} s^{-1} × 10^{10}).

spectrum of i-butane (see Chapter 4, Section II.A). They also are of importance in the radiolysis of hydrocarbons and have been studied extensively in radiolytic and mass spectrometric experiments by Ausloos and co-workers.[23,77,78]

Table 6 presents a summary of rate constants for hydride ion abstraction from alkanes by various reactant ions. The majority of these results come from mass spectrometric experiments[79,80] although some of the results for $t\text{-}C_4H_9^+$ come from radiolysis experiments.[23,81,82] In a few cases, there are differences of up to a factor of two in reported rate constants for the same reaction. In all cases, hydride ion transfer is the only reaction observed and calculated collision rate constants are in the range 13 to 16×10^{-10} cm^3 molecule^{-1} s^{-1}. The $C_2H_5^+$ and $s\text{-}C_3H_7^+$ ions react with the higher n-alkanes and branched alkanes with rate constants approaching the calculated collision rate constants. By contrast, the $t\text{-}C_4H_9^+$ ion is essentially unreactive with n-alkanes and shows a low reactivity (<6% of collision rate) with branched alkanes. The NO^+ ion shows a low reactivity with n-alkanes but a high reactivity, approaching the collision rate, with branched alkanes. The CF_3^+ ion is quite reactive with normal and branched alkanes showing rate constants approaching the calculated collision rate constants for the higher alkanes.

Thermochemical effects appear to play a significant role in these reactions. Table 7 summarizes the thermochemistry for H$^-$ abstraction from specific (primary, secondary, tertiary) positions of n-butane and i-butane by the reactant ions of Table 6. Also included in the table are the % H$^-$ abstraction from the specific positions as established by deuterium labelling. Reactions of $C_2H_5^+$ or CF_3^+ to abstract H$^-$ from primary, secondary, or tertiary positions are all exothermic, and reaction of all sites is observed. On the other hand, reaction of $s\text{-}C_3H_7^+$ to abstract H$^-$ from a primary position is endothermic, and only abstraction from secondary or tertiary positions is observed. Within reasonable error limits, only abstraction of H$^-$ from tertiary positions is thermochemically permitted with NO^+ as a reactant; consequently, NO^+ reacts at a significant rate only with those alkanes possessing a tertiary hydrogen. (In contrast with these

TABLE 7
Thermochemistry of Hydride Abstraction From
n-C$_4$H$_{10}$ and i-C$_4$H$_{10}$

	ΔH_{react} (% of total abstraction)			
	$C_2H_5^+$	$C_3H_7^+$	CF_3^+	NO^+
n-C$_4$H$_{10}$(prim)	−8 (—)	+14 (0)	−35 (36)	+18 (—)
n-C$_4$H$_{10}$(sec)	−26 (—)	−4 (100)	−53 (64)	0 (—)
i-C$_4$H$_{10}$(prim)	−8 (68)	+14 (0)	−35 (76)	+8 (0)
i-C$_4$H$_{10}$(tert)	−41 (32)	−19 (100)	−68 (24)	−15 (100)

Note: Thermochemical data from Table 8, % abstraction from Lias et al.[80]

results, NO CI of n-alkanes is reported[83,84] to produce [M–H]$^+$, although the sensitivity of the method has not been noted). The reaction of t-C$_4$H$_9^+$ to abstract the tertiary hydrogen from i-butane is thermoneutral while reaction with more complex branched olefins is only slightly exothermic. Consequently, it is not surprising that t-C$_4$H$_9^+$ reacts relatively slowly; detailed consideration[80] of the results also suggests that steric effects may play a role in the reactions of t-C$_4$H$_9^+$ while, for the remaining systems, there is some evidence for a dependence of the reaction rate on the lifetime of the ion/neutral complex.

Although quantitative data are restricted to the alkanes, hydride abstraction from other classes of molecules is known. For example, in chemical ionization studies, hydride ion abstraction from alcohols and aldehydes by NO$^+$ has been observed.[84,85] In fact, hydride ion abstraction is a possible reaction whenever Reaction 23 is exothermic, i.e., whenever the hydride ion affinity (HIA) of X$^+$ is greater than the HIA of [M – H]$^+$. HIAs are discussed in Section VI.B and a tabulation of HIAs is given in Table 8.

Olefinic ions also abstract H$^-$ from alkanes, but usually also react by H$_2^-$ abstraction, thus converting the olefinic ion to a neutral saturated hydrocarbon.[86–90] Thus, for example, the deuterated propene molecular ion reacts with isopentane by both Reactions 24 and 25

$$CD_3CDCD_2^{+\cdot} + i - C_5H_{12} \rightarrow CD_3CDCD_2H + C_5H_{11}^+ \qquad (24)$$

$$\rightarrow CD_3CDHCD_2H + C_5H_{10}^{+\cdot} \qquad (25)$$

with the ratio $k_{24}/k_{25} = 0.55$. The H$_2^-$ transfer reaction has found no practical application in chemical ionization mass spectrometry. Halide ion transfer reactions (Reaction 26, Y = halogen) also have

$$R_1^+ + R_2Y \rightarrow R_2^+ + R_1Y \qquad (26)$$

been observed[91–94] and may be of importance in some chemical ionization systems although the rates tend to be low.

TABLE 8
Hydride Ion Affinities of Gaseous Ions

R^+	$\Delta H_f^o(R^+)^a$	$\Delta H_f^o((RH)^b$	$HIA(R^+)$
CH_3^+	261	−17.9	313
H_3^+	265	0(2H_2)	300
CF_3^+	95	−166.3	296
$C_2H_5^+$	216	−20.2	271
CH_5^+	216	−17.9(H_2+CH_4)	269
NO^+	235	24^c	246
n-$C_3H_7^+$	211	−24.8	270
i-$C_3H_7^+$	191	−24.8	250
n-$C_4H_9^+$	203	−30.2	268
i-$C_4H_9^+$	199	−32.2	266
s-$C_4H_9^+$	183	−30.2	248
t-$C_4H_9^+$	166	−32.2	233
CH_2=CH–CH_2^+	226	4.9	256
CH_2=CH–C^+HCH_3	202	−0.03	237
$C_6H_5CH_2^+$	215	12.0	238
CH_3CH=OH^+	139	−56	230
$(CH_3)_2C$=OH^+	117	−65	217

Note: All data in kcal mol^{-1}.

^{a} From Lias et al.[154]
^{b} From Cox and Pilcher[166] unless otherwise indicated.
^{c} From Stull and Prophet.[167]

D. Condensation Reactions

In a condensation ion/molecule reaction, a relatively long-lived collision complex is formed involving the two reactants. This complex, under suitable conditions, may be collisionally stabilized, (Reaction 27) but in the absence of such stabilization decomposes, at least in part, to products different from the reactants (Reaction 28). This type of reaction is particularly prevalent in the

$$A^+ + B \rightarrow [AB^+]^* \xrightarrow{M} AB^+ \qquad (27)$$

$$\rightarrow C^+ + D \qquad (28)$$

gas-phase ion chemistry of alkenes and alkynes, involving reactions of both the molecular ion and fragment ions with the neutral, and normally leading to product ions of higher carbon content than the reactant ion. The total rate of reaction and the distribution of products formed depend strongly on the identity of the reactants. As an example, Table 9 summarizes the product distribution from the reaction of the $C_4H_8^{+\cdot}$ molecular ions (formed from the respective neutral olefins) with the C_4H_8 olefins listed; also included in the table are the total rate constants for reaction. These are, in general, lower than the collision rate constant of ~1.3 × 10^{-9} cm^3 molecule^{-1} s^{-1}. In the isobutene system, the

TABLE 9
Ion/Molecule Reactions in C_4H_8 Olefins ($C_4H_8^+ + C_4H_8 \rightarrow$ Products)

Product	Yield (% of total reaction)			
	i-C_4H_8	1-C_4H_8	2-C_4H_8	c-$C_3H_5CH_3$
$C_4H_9^+$	91.7	8.3	3.3	<1.0
$C_5H_9^+$	—	4.3	—	8.8
$C_5H_{10}^+$	4.5	38.8	5.8	69.0
$C_5H_{11}^+$	—	4.2	3.5	—
$C_6H_{11}^+$	3.4	26.3	42.3	8.0
$C_6H_{12}^+$	—	16.1	45.1	7.6
k_{react} [a]	5.4	6.0	0.37	0.60

Note: Data from Abramson, F.B. and Futrell, J.H., *J. Phys. Chem.*, 72, 1994, 1968. With permission.

[a] In cm^3 molecule^{-1} s$^{-1} \times 10^{10}$

major part of the reaction occurs by H^+/H-atom transfer to yield the $C_4H_9^+$ product, but with the remaining systems a variety of C_5 and C_6 products are formed. In many systems at higher pressure the collision complex is stabilized by collision and is observed.

The H^+/H-atom transfer reaction appears to occur by a different mechanism than the condensation reactions leading to the products of higher carbon number. Deuterium labelling results[96-98] show that the former reaction occurs by a "direct" mechanism in which there is no isotopic interchange between the reacting partners while the condensation reactions occur through formation of a long-lived intimate complex in which isotopic interchange between the reaction partners does occur. As an example, in the reaction of $C_3H_6^+$ (propylene) with $CD_2=CDCD_3$ only $C_3H_6D^+$ and $C_3D_6H^+$ are observed as $C_3(H,D)_7^+$ products while all possible $C_5(H,D)_9^+$ ions from $C_5H_6D_3^+$ to $C_5H_3D_6^+$ are observed.[96] Evidence has been presented[95,97,100-105] that an increase in the energy (internal or kinetic) of the reactants increases the importance of the H^+/H-atom transfer reaction at the expense of the condensation reactions.

Considerable interest has centered on the structures of the long-lived complexes formed in condensation ion/molecule reactions.[23] Of particular relevance in the present context is the evidence for four-centred cyclic complexes in the reactions of propylene,[98] allene and propyne,[106] and the fluoroethylenes.[107] Similar four-centered complexes have been invoked to rationalize the results obtained using vinyl methyl ether as a CI reagent gas for the determination of double bond position in olefinic compounds.[108,109]

Condensation ion/molecule reactions play only a minor role in chemical ionization studies. The most familiar example is Reaction 29 forming the $C_2H_5^+$

$$CH_3^+ + CH_4 \rightarrow C_2H_5^+ + H_2 \qquad (29)$$

reactant ion in methane at high pressure. At low ion kinetic energies, beam experiments studying the $CD_3^+ + CH_4$ reaction have shown that a long-lived complex is formed in which interchange of H and D does occur;[60] however, the complex is not sufficiently long-lived for $C_2H_7^+$ to be collisionally stabilized. The reaction rate constant for Reaction 29 is $\sim 1.0 \times 10^{-9}$ cm^3 molecule^{-1} s^{-1} only slightly lower than the calculated collision rate constant of 1.22×10^{-9} cm^3 molecule^{-1} s^{-1}.[52] The collisionally stabilized complexes of ion/molecule reactions frequently are observed in CI experiments. Thus, the $[M+C_2H_5]^+$ and $[M+C_3H_5]^+$ ions observed in CH_4 CI mass spectra can be considered as examples of stabilized complexes of condensation ion/molecule reactions. In the absence of stabilization, fragmentation of the complex may occur; these fragmentation reactions are discussed in Chapter 4.

E. Clustering or Association Reactions

$$A^+ + B \xrightarrow{\;M\;} AB^+ \tag{30}$$

In the association or clustering Reaction 30, the collision of an ion with a molecule results in formation of a complex which is observed after collisional stabilization. However, in contrast to the condensation reactions discussed above, in the absence of collisional stabilization, the complex dissociates to reform the reactants rather than to form new products. A particularly common type of clustering reaction is the solvation of gas-phase ions by polar molecules.

The mechanism of cluster ion formation is believed[110] to be

$$A^+ + B \underset{k_d}{\overset{k_c}{\rightleftharpoons}} AB^{+*} \tag{31}$$

$$AB^{+*} + M \xrightarrow{\;k_s\;} AB^+ + M \tag{32}$$

which leads to the overall rate constant expression

$$k_f = k_c k_s [M] \big/ \big(k_d + k_s[M]\big) \tag{33}$$

At low pressures, $k_d \gg k_s[M]$ leading to $k_f = k_c k_s[M]/k_d$ and a third order rate constant

$$k_f^{(3)} = k_c k_s / k_d \tag{34}$$

while at high pressures $k_s[M] \gg k_d$ and the clustering reaction becomes second order with a limiting rate constant $k_f^{(2)} = k_c$. At the usual pressures in the range 0.3 to 1 torr used in chemical ionization sources, most clustering reactions will be in the third-order region, while at atmospheric pressure, chemical ionization clustering reactions will be in the limiting second-order regime. Table 10 records third-order rate constants for a selection of clustering reactions measured at $\sim$300 K. These third-order rate constants usually have a T^{-n} temperature dependence where n is in the range 3 to 5. The increase in the third-order

TABLE 10
Association Reactions of Positive Ions

Reaction	k (cm^6 molecule^{-2} s^{-1})	References
$Ar^+ + 2Ar \rightarrow Ar_2^+ + Ar$	3×10^{-31}	Kebarle[111]
$O_2^+ + 2O_2 \rightarrow O_4^+ + O_2$	3×10^{-30}	Durden et al.[112]
$H_3O^+ + H_2O + CH_4 \rightarrow H^+(H_2O)_2 + CH_4$	4×10^{-27}	Meot-Ner and Field[113]
$NH_4^+ + 2NH_3 \rightarrow H^+(NH_3)_2 + NH_3$	5×10^{-27}	Nielson et al.[114]
$CH_3NH_3^+ + 2CH_3NH_2 \rightarrow H^+(CH_3NH_2)_2 + CH_3NH_2$	1.5×10^{-25}	Nielson et al.[114]

clustering rate constant with increasing complexity of the reactants is attributed to an increase in the lifetime of the activated species AB^{+*} (a decrease in k_d); similarly, the decrease in the rate constant with increasing temperature reflects the effect of the thermal energy on the lifetime of AB^{+*}. A third-order clustering rate constant of 1×10^{-25} cm^6 molecule^{-2} s^{-1} (at 300 K) corresponds to an effective second-order rate constant of $\sim 2 \times 10^{-9}$ cm^3 molecule^{-1} s^{-1} at a source pressure of M of 1 torr. Clearly, clustering reactions can be of considerable importance under CI conditions. In atmospheric pressure chemical ionization (APCI), clustering reactions will be in the second-order region with limiting rate constants given by k_c, usually assumed to be the collision rate constant.

IV. NEGATIVE ION/MOLECULE REACTIONS

Reactions between negative ions and neutral molecules have been studied less extensively and systematically than positive ion/molecule reactions. In addition, many of the detailed studies which have been made have involved species of interest with regard to the ion chemistry of the earth's atmosphere;[115] such species normally are not of major interest in chemical ionization studies. The following reviews those aspects of negative ion/molecule chemistry which are relevant to chemical ionization. Although electron/molecule interactions fall outside the definition of chemical ionization, electron capture chemical ionization (ECCI) has become a very important analytical method which normally is included in any discussion of chemical ionization; therefore, electron/molecule interactions will be reviewed.

A. Electron/Molecule Interactions
Negative ions are produced as a result of electron/molecule interactions by three general processes depicted for the molecule MX.

1. Ion-pair formation

$$e + MX \rightarrow M^+ + X^- + e \tag{35}$$

2. Electron attachment

$$e + MX \rightarrow MX^- \tag{36}$$

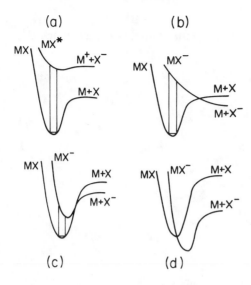

FIGURE 5.   Potential energy curves for negative ion formation.

3. Dissociative electron attachment

$$e + MX \rightarrow M + X^- \tag{37}$$

These processes may be better understood with reference to the potential energy diagrams depicted in Figure 5 for a hypothetical diatomic molecule; similar arguments will apply to polyatomic molecules if potential energy surfaces are considered. In the following discussion, the electron affinity (EA) is defined as the energy of the ground state neutral and an electron at infinite distance less the energy of the ground state of the negative ion.

In ion pair formation, Figure 5a, the electron provides the energy necessary to form an excited molecule which dissociates (or predissociates) to give a positive ion and a negative ion. The thresholds for such processes usually lie above 10 to 15 eV electron energy, and the cross section increases approximately linearly with excess energy to an energy roughly three times the threshold energy. Negative ion formation by ion-pair formation does not appear to be a particularly important process under CI conditions.

Dissociative electron attachment occurs (Figure 5b) if capture of an electron leads, by a vertical Frank-Condon transition, to a repulsive state of MX^- which dissociates to form $M \cdot$ and X^-, with possible excess internal and translational energy.

For electron attachment, two cases arise depending on whether the EA of MX is < 0 (Figure 5c) or > 0 (Figure 5d). When the EA is < 0, a Frank-Condon transition leads to an unstable molecular anion MX^{-*} which may disappear by autodetachment (Reaction 38) or, if above the dissociation limit, may

TABLE 11
Lifetimes of Molecular Anions

M⁻	t (μsec)
$C_6H_5NO_2^-$	~40
phthalic anhydride⁻	313
o-$ClC_6H_4NO_2^-$	17
o-$C_6H_4(NO_2)_2^-$	463
m-$NO_2C_6H_4COCH_3^-$	310
o-$CH_3OC_6H_4NO_2^-$	16
$C_6H_5F^-$	$<1 \times 10^{-6}$
$C_6F_6^-$	12
$C_6H_6^-$	1×10^{-6}
$C_6H_5CN^-$	5
$C_6F_5CN^-$	17
O_2^-	2×10^{-6}

Note: Data from Christophoru.[116]

$$MX^{-*} \rightarrow MX + e \qquad (38)$$

dissociate to M· + X⁻. Autodetachment, in this situation, usually is very rapid and occurs within a vibrational period leaving little possibility for collisional stabilization of MX^{-*} unless the pressure is very high. Stabilization by radiation emission also is possible but not probable.

When the EA of MX is greater than 0 (Figure 5d), the potential energy curve for MX⁻ may or may not intersect the vertical Frank-Condon region. If it does, the situation is similar to that discussed with respect to Figure 5c and similar reactions occur. If it does not, a different form of attachment is imagined. Essentially, because of the low (thermal) energy of the electron, long interaction times result. Under these conditions the Born-Oppenheimer approximation may break down and the nuclei relax and change to a position on the negative ion curve. If the excess energy of MX^{-*} can be distributed into the vibrational modes of the anion, significant lifetimes (with respect to autodetachment) for the molecular anion may be observed and collisional stabilization can occur more readily. All of the electron attachment processes are resonance processes since no electron is produced to carry away the excess energy. They are observed from thermal up to <15 eV electron energy, with thermal energies usually corresponding to electron attachment and higher energies corresponding to dissociative electron attachment.

The molecular anions formed by electron attachment are inherently unstable with respect to autodetachment (Reaction 38) and are stabilized by radiation emission or by collisional deactivation. For significant collisional stabilization at pressures of ~1 torr (CI pressures), the natural lifetime of the MX^{-*} species formed by electron attachment must be >1 μsec. Table 11

TABLE 12
Electron Attachment Rate Constants

M	k (cm^3 molecule^{-1} s^{-1})
CH_3Cl	-5.6×10^{-11}
CH_2Cl_2	1.4×10^{-11}
$CHCl_3$	3.6×10^{-9}
CCl_4	2.5×10^{-7}
CH_3CCl_3	1.5×10^{-8}
$CH_2ClCHCl_2$	1.4×10^{-10}
CF_3Cl	4.3×10^{-10}
CF_3Br	1.2×10^{-8}
CF_2Br_2	2.4×10^{-7}
CF_4	7.1×10^{-13}
$C_6H_5NO_2$	5.8×10^{-10}
C_6F_6	9.0×10^{-8}
$C_6F_5CF_3$	2.2×10^{-7}
Azulene ($C_{10}H_8$)	2.6×10^{-8}

summarizes lifetimes measured[116] for a number of anions formed by electron attachment; such lifetimes range from $<10^{-12}$ s to $>10^{-4}$ s. Ions with long lifetimes are molecules with positive EAs (often conjugated systems bearing electron-attracting substituents), and often are large molecules where the excess energy of MX^{-*} can be distributed among the many degrees of freedom to delay autodetachment. In a number of cases, the lifetimes are significantly greater than 10 μsec, and molecular anions should be observable in low pressure mass spectrometry where collisional stabilization will be negligible; this has been found to be so for a variety of compounds using derivatives containing conjugated dicarbonyl units[117,118] or nitrophenyl groups.[119,120]

A wide range of electron attachment rate coefficients also has been reported. Table 12 summarizes data for a variety of simple halogenated alkanes[121] and some aromatic molecules.[116] For the halogenated alkanes, the negative ions observed are halide anions while for the aromatic molecules, the molecular anions are observed. Of particular note is the fact that a number of the attachment rate coefficients are in the range 10^{-8} to 10^{-7} cm^3 molecule^{-1} s^{-1}, considerably higher than the maximum collision rate coefficients for ion/molecule reactions; in terms of Equations 14 or 18, the high rate coefficients arise from the low reduced mass of the electron/molecule pair. As a result of these high attachment rate coefficients, in favorable circumstances, electron capture chemical ionization (ECCI) can have a considerably higher sensitivity than chemical ionization techniques relying on ion/molecule reactions. This aspect will be discussed further in Chapter 4.

B. Associative Detachment Reactions
Negative ions may disappear from the system by the associative detachment reaction

TABLE 13
Rate Constants for Displacement Reactions $X^- + RY \rightarrow RX + Y^-$

X^-	RY	k $(cm^3\ molecule^{-1}\ s^{-1} \times 10^9)$	Reaction efficiency
OH^-	CH_3Cl	2.0	0.84
F^-	CH_3Cl	1.3	0.56
CH_3O^-	CH_3Cl	1.3	0.65
$CF_3CH_2O^-$	CH_3Cl	0.22	0.15
SH^-	CH_3Cl	0.012	0.006
OH^-	CH_3Br	2.2	0.95
F^-	CH_3Br	2.0	0.88
CH_3O^-	CH_3Br	1.7	0.93
$CF_3CH_2O^-$	CH_3Br	0.89	0.70
SH^-	CH_3Br	0.32	0.17

$$X^- + N \rightarrow XN + e \qquad (39)$$

where N is a neutral species. Such reactions have been studied primarily in simple systems. For example, the O^- ion has been found[122] to react with C_2H_2 at 494 K with a total rate constant of 2.1×10^{-9} cm^3 $molecule^{-1}$ s^{-1} to give $C_2H_2O + e$ (45%), $C_2^- + H_2O$ (35%), $C_2H^- + HO$ (46%), and C_2HO^- (6%). The prevalence of associative detachment reactions in the more complex systems applicable to chemical ionization studies has not been studied, however, it is a complication which should be borne in mind since it can lead to a decrease in sensitivity.

C. Displacement and Elimination Reactions

The displacement reaction

$$X^- + RY \rightarrow RX + Y^- \qquad (40)$$

is the gas phase equivalent of the S_N2 reaction so extensively studied in solution. Lieder and Brauman,[123,124] in a study of the reaction of Cl^- with *cis*- and *trans*-4-bromocyclohexanol in the gas-phase, have shown that the reaction occurs with inversion of configuration as expected for a true S_N2 displacement reaction. Displacement reactions generally are observed if they are exothermic and if there is no facile proton transfer process from RY to X^-. For example, the OH^- ion reacts with CH_3CN by proton abstraction rather than displacement of CN^-.[73]

Table 13 presents a summary of rate constants for some simple displacement reactions taken from the recent work of DePuy et al.;[125] also given in the table are the reaction efficiencies defined as k_{react}/k_{collis}, with k_{collis} calculated by the ADO theory. A wide range of reaction efficiencies is observed in this and other work;[126] these efficiencies do not correlate with the reaction thermo-

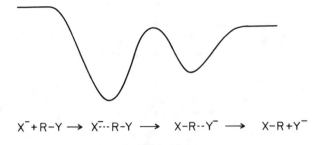

$$X^- + R-Y \longrightarrow X^-\cdots R-Y \longrightarrow X-R\cdots Y^- \longrightarrow X-R+Y^-$$

FIGURE 6.   Potential energy surface for displacement reaction.

chemistry but rather depend in detail on the specific nucleophile X^- and leaving group Y^-. The results have been rationalized by Brauman and colleagues[126,127] in terms of a potential energy surface for the reaction involving two minima as illustrated in Figure 6. The height of the barrier between the two minima strongly influences the rate of the reaction. A similar potential energy surface is involved in proton transfer reactions of positive or negative ions. The data in Table 13 involve only methyl halides. For higher alkyl halides DePuy et al.[125] have concluded that oxyanions react almost exclusively by elimination, Reaction 41, (illustrated by n-propyl bromide), rather than by displacement. By

$$RO^- + CH_3CH_2CH_2Br \rightarrow ROH + CH_3CH = CH_2 + Br^- \tag{41}$$

contrast, they concluded that sulfur anions reacted inefficiently by displacement with the higher alkyl halides. From analysis of neutral products, Jones and Ellison[128] also concluded that CH_3O^- reacted with n-propyl bromide largely, if not exclusively, by elimination.

Displacement or elimination reactions are of minimal utility in CIMS and could be a potential nuisance since they provide little structural information. One example where a displacement reaction is known to be involved is in the formation of carboxylate anions by reaction of OH^- or O^- with esters. Isotopic labelling experiments[129] have shown that both an S_N2 displacement and a direct attack at the carbonyl oxygen are involved.

D. Proton Transfer Reactions

The proton transfer reaction

$$X^- + YH \rightarrow XH + Y^- \tag{42}$$

will be exothermic if the gas-phase acidity of YH is greater than the gas-phase acidity of XH. The gas-phase acidity scale is defined (Section V.D) in terms of the ΔH^0 of the reaction

$$XH \rightarrow X^- + H^+ \quad \Delta H^0 = \Delta H_{acid} \tag{43}$$

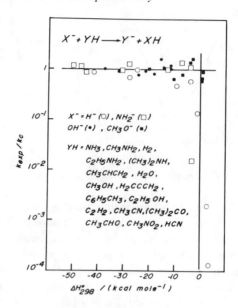

FIGURE 7. Efficiency of proton abstraction as a function of ΔH^0. (From Bohme, D.K., *Trans. Roy. Soc. Can., Series IV*, XIX, 265, 1981. With permission.)

Since Reaction 43 is endothermic, the acidity is greater the smaller the ΔH_{acid}. Alternatively, the enthalpy change for Reaction 43 represents the PA of X^-, and an equivalent statement is that Reaction 42 will be exothermic if the PA of X^- is greater than the PA of Y^-.

The proton transfer reactions of negative ions have been studied extensively under equilibrium conditions to establish relative gas-phase acidities,[16,130] however, there have been fewer kinetic studies. Figure 7 summarizes results obtained by Bohme and co-workers[131] for reactions of H^-, NH_2^-, OH^- and CH_3O^- with a variety of simple molecules plotted as the reaction efficiency (k_{exp}/k_{collis}) versus ΔH^0 for the reaction where k_{collis} is the collision rate constant calculated from ADO theory. As shown, these reactions show unit efficiency provided they are at least 10 kcal mol^{-1} exothermic. The reaction efficiencies tend to drop off as the reaction approaches thermoneutrality and become low for endothermic reactions. For reactions involving delocalized anions, such as $CH_3COCH_2^-$, rather low reaction efficiencies are observed.[132]

Proton transfer reactions are of importance in chemical ionization studies not only because of their generally high efficiency but also because they lead to formation of $[M-H]^-$ ions which provide molecular mass information. In particular, the anion OH^- has seen considerable use; since the PA of OH^- is quite high, it tends to react readily by proton abstraction from many organic molecules.

TABLE 14
Charge Exchange Reactions of Negative Ions

Reaction	k (cm^3 molecule^{-1} s^{-1} × 10^{10})	References
$O^- + NO_2 \rightarrow NO_2^- + O$	12	Ferguson et al.[133]
$O_2^- + NO_2 \rightarrow NO_2^- + O_2$	8	Ferguson et al.[133]
$O_2^- + O_3 \rightarrow O_3^- + O_2$	8	Fahey et al.[134]
$O^- + SF_4 \rightarrow SF_4^- + O$	14	Babcock and Streit[135]
$S^- + SF_4 \rightarrow SF_4^- + S$	11	Babcock and Streit[135]

E. Charge Exchange Reactions

The charge exchange reaction

$$A^- + B \rightarrow B^- + A \qquad (44)$$

will be exothermic provided the EA of B is greater than the EA of A. The rate constants for a number of simple exothermic negative ion charge exchange reactions are summarized in Table 14. These results indicate reaction efficiencies in the range 30 to 100%. More recently, Kebarle and Chowdhury[136] have reported measurement of rate constants for exothermic charge exchange reactions involving a variety of nitroaromatic molecules as well as NO_2; the experimental rate constants were close to those calculated by ADO theory. By contrast, much lower rates were observed for reactions involving SF_6, perfluorocyclohexane, and perfluoromethylcyclohexane anions reacting with nitroaromatic molecules. It was concluded[136] that these reactions were inefficient because the geometries of the neutral perfluoro compound and its negative ion are quite different.

F. Association Reactions

Third-order association or clustering reactions are as common for negative ions as for positive ions. Thus, anions in the presence of polar molecules such as water form solvated species. The mechanism of such clustering reactions will be the same as outlined in Reactions 31 and 32 for positive ions. Although the equilibria involved in such clustering reactions have been extensively studied,[16,111] the kinetic parameters have not been as extensively studied. Table 15 reports third-order rate constants for a number of association or clustering reactions for relatively simple systems. There is no reason not to believe that for more complex systems the clustering reactions of negative ions will be as efficient as the clustering reactions of positive ions. Thus, from the data of Table 10, one could expect that for complex systems association reactions of negative ions could have third-order rate coefficients of 10^{-25} cm^6 molecule^{-2} s^{-1}, corresponding to an effective second-order rate constant at 1 torr pressure of about 10^{-9} cm^3 molecule^{-1} s^{-1}.

TABLE 15
Association Reactions of Negative Ions

Reaction	k $(cm^6 molecule^{-2} s^{-1})$	References
$O^- + 2O_2 \rightarrow O_3^- + O_2$	9×10^{-31}	Beaty et al.[140]
$Cl^- + H_2O + O_2 \rightarrow Cl^-(H_2O) + O_2$	2×10^{-29}	Fehsenfeld and Ferguson[137]
$NO_3^- + H_2O + O_2 \rightarrow NO_3^-(H_2O) + O_2$	7.5×10^{-29}	Payzant et al.[138]
$HO^- + H_2O + O_2 \rightarrow HO^-(H_2O) + O_2$	2.5×10^{-28}	Fehsenfeld and Ferguson[137]
$CF_3^- + CO_2 + He \rightarrow CF_3^-(CO_2) + He$	1.1×10^{-27}	McDonald and Chowdhury[139]

V. THERMOCHEMICAL PROPERTIES OF GAS-PHASE IONS

In this section, we review the thermochemical properties of gas-phase ions which are of relevance to CIMS. Many of these thermochemical results have been derived from equilibrium studies of gas-phase ion/molecule reactions

A. Gas-Phase Basicities: Proton Affinities (PAs)

The fundamental concept of basicity, defined by Brønsted[141] as the tendency of a molecule to accept a proton in the reaction

$$B + H^+ \rightarrow BH^+ \qquad (45)$$

has long been of concern to chemists. By definition, $-\Delta H^0$ for Reaction 45 is designated as the PA of the base B. Although relative basicities in solution have been established for some time, it is only relatively recently that gas-phase basicities, free of the complication of solvation, have been measured by mass spectrometric techniques. Early gas-phase measurements used reaction bracketing techniques[142,143] to establish the direction of exothermic proton transfer between base pairs, thus providing approximate relative PAs. In effect, rapid proton transfer reactions were assumed to be exothermic and slow reactions were assumed to be endothermic. More recently, quantitative measurements of equilibrium constants for reactions of the type

$$B_1H^+ + B2 \rightleftharpoons B_2H^+ + B_1 \qquad (46)$$

by ion cyclotron resonance techniques,[144] by high pressure mass spectrometry,[145] and by flowing afterglow experiments[146] have permitted the determination of relative PAs to a precision of ~0.2 kcal mol^{-1}. By the use of suitable reference compounds, whose PAs are known from other measurements, these relative values can be put on an absolute scale. Many of these measurements have been made at a single temperature and, thus, yield ΔG^0 for Reaction 46. To establish ΔH^0, the difference in PAs, the standard entropy change, ΔS^0, must be evaluated. If the temperature dependence of the equilibrium constant is measured experimentally, both ΔH^0 and ΔS^0 can be evaluated; in the absence of such variable temperature measurements it usually has been assumed that

TABLE 16
Proton Affinities of Substances Less Basic Than H_2O

B	$PA(B)^a$ (kcal mol^{-1})
O_2	101.5
H_2	101.2
Kr	101.6
N_2	118.2
Xe	118.6
CO_2	130.8
CH_4	131.6
N_2O	138.8
CO	142.4

[a] From Lias et al.[154]

the only contribution to ΔS^0 arises from symmetry changes, and calculations are made on the basis of this assumption. The majority of results support this assumption although exceptions have been noted.[147]

Relative PAs also can be evaluated from studies of the fragmentation (unimolecular or collision-induced) of proton-bound dimers $B_1 \cdots H^+ \cdots B_2$.[148] These loosely bound ions usually fragment by either or both of Reactions 47

$$B_1 \cdots H^+ \cdots B_2 \rightarrow B_1 H^+ + B_2 \qquad (47)$$

$$\rightarrow B_2 H^+ + B_1 \qquad (48)$$

and 48 and it has been shown[148] that the relative abundances of the two products can be correlated with the relative PAs of the two bases B_1 and B_2. This method has been used recently to obtain the relative PAs of amino acids.[149]

A number of extensive reviews and tabulations of gas-phase PAs are available.[150-154] In these tabulations, the absolute values of PAs, but not the relative values, show some differences because the absolute assignment of the PA scale is still in some doubt and the different authors have chosen different standards to establish the absolute scale. Tables 16 to 20 summarize a large body of PA data; all data are taken from the most recent tabulation.[154] More complete tabulations are available elsewhere.[151,152,154]

A detailed evaluation of the PA data and the variation of PA with structure has been presented elsewhere[151] and need not be repeated here. For the present purposes, it is sufficient to note that oxygen-containing compounds have PAs in the 180 to 205 kcal mol^{-1} range, sulfur-containing compounds in the 198 to 209 kcal mol^{-1} range, and nitrogen-containing compounds in the 205 to 240 kcal mol^{-1} range. Benzenoid compounds have PAs in the 182 to 212 kcal mol^{-1} range depending on the substituent. Thus, it is clear that BH^+ ions derived from any

TABLE 17
Proton Affinities of Alkenes and Alkadienes

Compounds	PA[a] (kcal mol^{-1})
$CH_2=CH_2$	162.6
$CH_2=CH-CH_3$	179.5
$CH_3CH=CHCH_3$	180.3
$CH_2=C(CH_3)CH_3$	195.9
$CH_2=CH-CH=CH_2$	190
$CH_2=CH-CH=CHCH_3$	201.8
cyclo–C_5H_6	199.6

[a]   Calculated from data in Lias et al.[154]

of the bases listed in Table 16, as well as the $C_2H_5^+$ ions also produced in high pressure methane, will protonate all organic bases in an exothermic reaction. On the other hand, protonation of some organic bases by CI reagents such as $C_4H_9^+$ (protonated isobutene) or NH_4^+ will be endothermic and, thus, rather slow. In these cases, other reaction modes, such as clustering, may become important; these aspects are discussed in more detail in Chapter 4.

The PA data in the tables refer to protonation at the most basic site of the molecule, and for those molecules with more than one basic site, the question arises as to the site of protonation. For simple carboxylic acids and esters, correlations of the PA data with O(1S) ionization energies have established[155-157] that the compounds are protonated at the carbonyl oxygen; the PAs of the singly bonded oxygens are 18 to 25 kcal mol^{-1} lower.[157] Similar results indicate that amides are preferentially protonated at the oxygen.[158] For substituted benzenes, protonation of the aromatic ring occurs except for those with NO_2, CHO, and CN substituents where protonation occurs at the substituent.[159] These results refer to protonation of the substrate under conditions of thermodynamic control; by contrast, protonation reactions under chemical ionization conditions may be under kinetic control and the kinetically favored site of protonation may differ from the thermodynamically favored site. Consequently, it is not safe to assume that, in chemical ionization, protonation occurs only at the thermodynamically favored site. For example, it has been shown[160,161] that, in the chemical ionization of halobenzenes, significant protonation occurs at the halogen even though protonation of the aromatic ring is thermodynamically favored. In many cases, the proton may exchange between the available basic sites in the molecule; this has been shown to be the case in the protonation of amino acids under CI conditions.[162]

Very recently, Meot-Ner and Sieck[163] have published evidence that the PA scale between i-butene and trimethyl amine should be expanded in a linear fashion. Thus, while the PA of i-butene remains 195.9 kcal mol^{-1} that for trimethyl amine increases from 225.1 to 232.7 kcal mol^{-1}, with compounds in between having a proportional increase in PA. This publication clearly illustrates the problems associated with establishing absolute PAs, however, the

TABLE 18
Proton Affinities of Substituted Benzenes

Substituent	PA[a] (kcal mol^{-1})
H	181.3
Cl	181.7
F	182.6
CH$_3$	189.8
C$_2$H$_5$	191.6
n-C$_3$H$_7$	192.4
i-C$_3$H$_7$	192.1
t-C$_4$H$_9$	193.0
NO$_2$	193.4
OH	196.3
CN	195.9
CHO	200.2
OCH$_3$	200.3
NH$_2$	209.5

[a] From Lias et al.[154]

changes proposed do not affect the arguments presented here since they are based more on relative PAs.

B. Hydride Ion Affinities (HIAs)

The HIA of a gaseous cation is defined as $-\Delta H^0$ for the reaction

$$R^+ + H^- \rightarrow RH \qquad -\Delta H^0 = HIA(R^+) \qquad (49)$$

Since $\Delta H_f^o(H^-) = 34.6$ kcal mol^{-1},[154] HIAs can be calculated provided the heats of formation of R$^+$ and RH are known. Relative HIAs can be obtained, in favorable cases, from studies[164,165] of the equilibrium

$$R_1^+ + R_2H \rightleftharpoons R_2^+ + R_1H \qquad (50)$$

A partial list of HIAs is given in Table 8. A more complete listing can be found elsewhere.[151]

The HIA scale is probably the most useful and convenient basis for the comparison of the relative stabilities of carbocations. Of more relevance in the present context is the ordering of cations with respect to the ability to abstract H$^-$ in chemical ionization reactions. The hydride ion abstraction Reaction 50 will be exothermic provided HIA(R$_1^+$) > HIA(R$_2^+$). Accordingly, CH$_3^+$, CF$_3^+$ and H$_3^+$, with high HIAs, should be capable of abstracting H$^-$ from most organic substrates. The CH$_5^+$ and C$_2$H$_5^+$ ions are slightly weaker H$^-$-abstracting reagent ions while NO$^+$ is a relatively weak H$^-$-abstracting reagent. The t-C$_4$H$_9^+$ ion is a very weak H$^-$-abstracting reagent ion. Of course a number of these species are Brønsted acids which also can react by proton transfer; the competition between these reaction modes will be discussed in more detail in Chapter 4.

TABLE 19
Proton Affinities Between H_2O and NH_3

Alcohols and ethers		Aldehydes and ketones		Acids and esters		Thiols, sulphides, and nitriles	
H_2O	166.5	HCHO	171.7	CF_3CO_2H	169.0	H_2S	170.2
CF_3CH_2OH	169.0	MeCHO	186.6	HCO_2H	178.8	MeSH	187.4
MeOH	181.9	EtCHO	189.6	$MeCO_2H$	190.2	EtSH	190.8
EtOH	188.3	n-PrCHO	191.5	$EtCO_2H$	191.8	i-PrSH	194.1
n-PrOH	190.8	n-BuCHO	192.6	HCO_2Me	188.4	Me_2S	200.6
i-PrOH	191.2			HCO_2Et	193.1	Et_2S	205.1
n-BuOH	191.1			HCO_2n-Pr	194.2	i-Pr_2S	209.6
t-BuOH	193.7	Me_2CO	196.7	$MeCO_2Me$	197.8	HCN	171.0
Me_2O	192.1	MeCOEt	199.8	$MeCO_2Et$	200.7	MeCN	188.2
Et_2O	200.2	Et_2CO	201.4	$EtCO_2Me$	200.2	EtCN	192.6
n-Pr_2O	202.3					n-PrCN	193.7
n-Bu_2O	203.7						
THF	198.8						
THP	203.6						

Note: PAs (in kcal mol^{-1}) from Lias et al.[154]

C. Electron Affinities (EAs)

A fundamental property of a negative ion is the lowest energy required to remove an electron. This energy is defined as the EA of the anion. For the reaction

$$X^- \rightarrow X + e \qquad \Delta H^0 = EA\left(X^-\right) \qquad (51)$$

ΔH^0 = EA provided X^- and X are in their ground rotational, vibrational, and electronic states and provided the electron has zero kinetic and potential energy. If one wishes to focus on the neutral species X, the standard enthalpy change for Reaction 51 can be defined as the affinity of X for an electron; this is an entirely equivalent definition.

There are two general types of experiments that have been employed successfully in determining absolute EAs. In optical methods, photons interact with negative ions to remove the least tightly bound electron; the energy required to remove this electron is equated with the EA. In laser photoelectron spectroscopy,[168] a fixed frequency laser is employed and the energy of the detached electrons are analyzed. In laser photodetachment spectroscopy[169,170] the laser wavelength is varied to determine the threshold for detachment of a (presumably) thermal electron. These optical studies often have involved even electron negative ions such as enolate and alkoxide ions. More recently, Kebarle and Chowdhury[136] have determined, by high pressure mass spectrometry, equilibrium constants for electron transfer reactions of odd-electron molecular anions (Reaction 52). From the

TABLE 20
Proton Affinities Greater Than NH₃

Amines		α, φ-diamines		Pyridines		Anilines	
NH_3	204.0	$H_2N(CH_2)_2NH_2$	225.9	H	220.3	H	209.5
$MeNH_2$	214.1	$H_2N(CH_2)_3NH_2$	234.1	3-Me	224	m-Me	213.4
$EtNH_2$	217.0	$H_2N(CH_2)_4NH_2$	237.6	4-Me	225.2	p-Me	213.7
$n\text{-}PrNH_2$	217.9	$H_2N(CH_2)_5NH_2$	238.1	2,4-Me₂	227.3	m-MeO	217.6
$i\text{-}PrNH_2$	218.6	$Me_2N(CH_2)_4NMe_2$	246.0	3-CN	209.3	m-Cl	207.2
$n\text{-}BuNH_2$	218.4	$Me_2N(CH_2)_6NMe_2$	245.0	3-F	214.3	p-Cl	208.6
Me_2NH	220.6			3-Cl	214.8	m-F	208.1
Me_3N	225.1			4-Cl	217.8	N-Me	218.1
Et_2NH	225.9			4-MeO	227.6	N-Me₂	223.4
Et_3N	232.3			3-NH₂	221.0		
$n\text{-}Pr_2NH$	227.5			4-NO₂	208.5		

Note: PAs (in kcal mol⁻¹) from Lias et al.[154]

$$M_1^- + M_2 \rightleftharpoons M_2^- + M_1 \tag{52}$$

temperature dependence of the equilibrium constant, one can derive both ΔH^0 and ΔS^0, thus establishing relative EAs. By constructing a ladder of relative EAs which includes compounds of known EA, absolute EAs can be determined. EAs also have been determined by studying the endothermic charge transfer reaction

$$X^- + Y \rightarrow Y^- + X \tag{53}$$

as a function of the translational energy of X^-. From the observed translational energy threshold, the endothermicity can be estimated and the EA of Y^- estimated if the EA of X^- is known.[171] This method gives only lower estimates for the EA since Y^- is not necessarily formed in its ground state.

Table 21 presents a listing of selected EAs taken from the recent compilation of thermochemical data.[154] These data have been discussed in more detail in other publications.[136,169,170,173]

Benzene has a negative EA but hexafluorobenzene and nitrobenzene have positive EAs illustrating the effect of adding electronegative groups to the aromatic system. The usual negative ion chemical ionization reagent ions O^-, Cl^-, OH^-, and CH_3O^- have large EAs and are unlikely to undergo exothermic charge exchange reactions with most organic molecules. On the other hand, O_2^- has a low EA and may well react by charge exchange.

D. Gas-Phase Acidities: PAs of Anions

The gas-phase acidity scale is defined in terms of the ΔH^0 for the gas-phase reaction

TABLE 21
Table of Electron Affinities

X⁻	EA(X⁻) (kcal mol⁻¹)	References
H⁻	17.4	Mead et al.[170]
O⁻	33.7	Mead et al.[170]
F⁻	78.4	Mead et al.[170]
Cl⁻	83.4	Mead et al.[170]
Br⁻	77.6	Mead et al.[170]
OH⁻	42.2	Drzaic et al.[169]
O_2^-	10.1	Drzaic et al.[169]
CH_3O^-	36.2	Drzaic et al.[169]
CH_3S^-	43.4	Drzaic et al.[169]
$C_3H_5^-$	12.7	Drzaic et al.[169]
$c-C_5H_5^-$	41.2	Drzaic et al.[169]
$C_6H_6^-$	<0	Christophoru and Goans[172]
$C_6F_6^-$	12.0	Kebarle and Chowdhury[136]
$C_6F_5CF_3^-$	21.7	Kebarle and Chowdhury[136]
$C_6H_5NO_2^-$	23.3	Kebarle and Chowdhury[136]
$p-NO_2C_6H_4NO_2^-$	46.1	Kebarle and Chowdhury[136]
Azulene $(C_{10}H_8^-)$	15.9	Kebarle and Chowdhury[136]
Phthalic anhydride⁻	27.9	Kebarle and Chowdhury[136]

$$XH \rightarrow X^- + H^+ \quad \Delta H^0 = \Delta H^0_{acid} \qquad (54)$$

Since Reaction 54 is invariably endothermic, the gas-phase acidity decreases as ΔH^0_{acid} increases. Alternatively, ΔH^0_{54} can be considered as the PA of the anion X^-. In some cases, absolute acidities can be evaluated from bond dissociation energies and EAs using the following thermochemical cycle:

$$XH \rightarrow X + H \qquad \Delta H^0 = D(X-H) \qquad (55)$$

$$X + H \rightarrow X^- + H^+ \qquad \Delta H^0 = IE(H) - EA(X) \qquad (56)$$

$$XH \rightarrow X^- + H^+ \quad \Delta H^0_{acid} = D(X-H) + IE(H) - EA(X) \qquad (57)$$

Since the IE of H is known (IE[H] = 313.6 kcal mol⁻¹ [154]), the absolute acidity of HX can be evaluated if the bond dissociation energy, D(H–X), and the EA, EA(X), are known. One might also note that measurement of gas-phase acidities and EAs has become a powerful method for determining bond dissociation energies through Equation 57; see, for example, the recent determination of the bond energies in acetylene and ethylene.[174] If absolute acidities cannot be evaluated by Equation 57, relative acidities can be established from measurements of the equilibrium constants for the proton transfer reaction

TABLE 22
ΔH^0_{acid} For Simple Brønsted Acids (HX $\rightleftharpoons$ H$^+$ + X$^-$)

HX	ΔH^0_{acid} (kcal mol^{-1})
H$_2$	400.3
NH$_3$	403.7
H$_2$O	390.8
OH	382.2
CH$_3$OH	380.5
HF	371.4
HO$_2$	352.7
HCN	351.1
HCl	333.4
HBr	323.6

Note: Data from Lias et al.[154]

$$X_1^- + X_2H \rightleftharpoons X_2^- + X_1H \tag{58}$$

either by measuring the equilibrium constant directly or by determining the rate constants for the forward and reverse reactions. These relative acidities can be put on an absolute scale by including in the measurements an acid whose acidity has been derived by the above thermochemical cycle. Relative acidities also have been evaluated by reaction bracketing, i.e., by observation of the rates of proton transfer reactions, it being assumed that endothermic reactions will be very slow while exothermic reactions will occur at an easily measured rate. Equilibrium measurements at a single temperature lead directly to ΔG^0 values; to derive ΔH^0 values the standard entropy change must be evaluated. The entropy change normally has been derived by assuming that the only significant contributions to ΔS^0 are rotational entropy changes arising from symmetry changes and the freezing out or loss of internal rotations. Measurements of the temperature dependence of the equilibrium constant lead to ΔH^0 and ΔS^0 directly and support the above assumption. Relative gas-phase acidities also can be obtained from studies of the fragmentation (either unimolecular or collision-induced) of proton-bound cluster ions, Reactions 59 and 60. It has been shown[175,176] that the ratio of intensities of the two anions X_1^- and X_2^- can be correlated with the relative gas phase acidities of X_1H and X_2H.

$$\left[X_1 \cdots H \cdots X_2\right]^- \rightarrow X_1^- + X_2H \tag{59}$$

$$\rightarrow X_2^- + X_1H \tag{60}$$

The various approaches to the measurement of gas-phase acidities have been reviewed.[130,154]

TABLE 23
ΔH^0_{acid} FOR ORGANIC COMPOUNDS

Alcohols		Phenols		Carboxylic acids		Carbonyl compounds	
MeOH	380.5	H	349.2	$MeCO_2H$	348.2	CH_3CHO	365.9
EtOH	377.4	m-Me	349.7	$EtCO_2H$	347.5	$MeCH_2CHO$	365.2
n-PrOH	376.0	m-MeO	348.0	HCO_2H	345.2	$(CH_3)_2CO$	369.0
i-PrOH	375.5	m-NH_2	350.6	$PhCO_2H$	338.9	$PhCOCH_3$	361.4
i-BuOH	374.8	m-F	343.7	FCH_2CO_2H	338.4	$PhCH_2COCH_3$	350.1
t-BuOH	374.5	m-Cl	342.0	$ClCH_2CO_2H$	336.3	CH_3CONMe_2	375.0
$PhCH_2OH$	370.0	p-Cl	343.2	CF_3CO_2H	322.9	CH_3COCF_3	350.4
F_2CHCH_2OH	366.4	m-CN	335.8			$PhCOCH_2COMe$	339.9
F_3CCH_2OH	361.9	p-NO_2	327.9			CH_3CO_2Me	371.9

Thiols		Acetylenes		Nitriles and nitro compounds		Hydrocarbons	
MeSH	356.8	$HC{\equiv}CH$	376.7	CH_3CN	372.8	C_6H_6	400.8
EtSH	355.2	$MeC{\equiv}CH$	381.2	$MeCH_2CN$	375.0	$c\text{-}C_5H_6$	354.0
n-PrSH	354.2	$nPrC{\equiv}CH$	379.8	$PhCH_2CN$	350.6	$c\text{-}C_7H_8$	375.2
i-PrSH	353.5	$t\text{-}BuC{\equiv}CH$	378.1	CH_3NO_2	356.4	$PhCH_3$	380.7
t-BuSH	352.5	$PhC{\equiv}CH$	370.7	$MeCH_2NO_2$	356.1	$PhCH_2Me$	379.8
		$CH_2{=}CH{-}CH_2$	380.5			Ph_2CH_2	363.5
						Fluorene	351.8
						$CH_3CH{=}CH_2$	390.8

Anilines	
H	366.4
p-MeO	367.1
m-Me	366.9
m-F	361.1
m-Cl	361.1
m-NO_2	352.3
p-NO_2	343.5

Note: Data (in kcal mol^{-1}) from Lias et al.[154]

Table 22 summarizes the gas-phase acidities of Brønsted acids, XH, where the anion X$^-$ has been, or could be, used as a proton abstraction chemical ionization reagent ion. Table 23 presents a selection of acidities for a variety of organic molecules; a more complete tabulation can be found elsewhere[154] as well as a discussion of structural effects on gas-phase acidities.[130,177] The compounds in Table 22 have been listed in order of increasing acidity or decreasing PA of the corresponding anion. The larger ΔH^0_{acid}(XH) the greater is the probability that proton abstraction by X$^-$ will be exothermic. Obviously anions such as NH_2^-, OH$^-$, or O$^-$ should be capable of exothermically abstracting a proton from a wide variety of molecules. On the other hand, O_2^-, Cl$^-$ and Br$^-$ are much weaker reagents and will show a limited capability for proton abstraction.

REFERENCES

1. **Dempster, A.J.**, The ionization and dissociation of hydrogen molecules and the formation of H_3, *Philos. Mag.*, 31, 438, 1916.
2. **Smyth, H.D.**, Primary and secondary products of ionization in hydrogen, *Phys. Rev.*, 25, 452, 1925.
3. **Hogness, T.R. and Lunn, E.G.**, The ionization of hydrogen as interpreted by positive ray analysis, *Phys. Rev.*, 26, 44, 1925.
4. **Hogness, T.R. and Harkness, R.W.**, The ionization of iodine interpreted by the mass spectrograph, *Phys. Rev.*, 32, 784, 1928.
5. **Smyth, H.D.**, Products and processes of ionization by low speed electrons, *Rev. Mod. Phys.*, 3, 347, 1931.
6. **Thompson, J.J.**, *Rays of Positive Electricity*, Longmans Green, London, 1931.
7. **Tal'roze, V.L. and Lyubimova, A.K.**, Secondary processes in a mass spectrometer ion source, *Dokl. Akad. Nauk. SSSR.*, 86, 969, 1952.
8. **Stevenson, D.P. and Schissler, D.O.**, Rate of the gaseous reaction $X^+ + YH \rightarrow XH^+ + Y$, *J. Chem. Phys.*, 23, 1353, 1955.
9. **Field, F.H., Franklin, J.L., and Lampe, F.W.**, Reactions of gaseous ions. I. Methane and ethylene, *J. Am. Chem. Soc.*, 79, 2419, 1957.
10. **Stevenson, D.P.**, Ion-molecule reactions, in *Mass Spectrometry*, McDowell, C.A., Ed., McGraw-Hill, New York, 1963.
11. **Melton, C.E.**, Ion-molecule reactions, in *Mass Spectrometry of Organic Ions*, McLafferty, F.W., Ed., Academic Press, New York, 1963.
12. **Henchman, M.J.**, Ion-molecule reactions and reactions in crossed molecular beams, *Annu. Rep. Chem. Soc.*, 62, 39, 1965.
13. **Tal'roze, V.L. and Karachevtsev, G.V.**, Ion-molecule reactions, *Adv. Mass Spectrom.*, 3, 211, 1966.
14. **Futrell, J.H. and Tiernan, T.O.**, Ion-molecule reactions, in *Fundamental Processes in Radiation Chemistry*, Ausloos, P., Ed., J. Wiley & Sons, New York. 1968.
15. **Ferguson, E.E.**, Ion-molecule reactions, *Annu. Rev. Phys. Chem.*, 26, 17, 1975.
16. **Kebarle, P.**, Ion thermochemistry and solvation from gas phase ion equilibria, *Annu. Rev. Phys. Chem.*, 28, 445, 1977.
17. **Jennings, K.R.**, Recent developments in the study of ion-molecule reactions, *Adv. Mass Spectrom.*, 7, 209, 1978.
18. **Jennings, K.R.**, Low energy ion-molecule reactions, in *Advances in Mass Spectrometry, 1985*, Todd, J.F.J., Ed., John Wiley & Sons, New York, 1986.
19. **Ausloos, P.**, Ed., *Ion Molecule Reactions in the Gas Phase*, Adv. Chem. Ser. 58, American Chemical Society, Washington, D.C., 1966.
20. **McDaniel, E.W., Cermak, V., Dalgarno, A., Ferguson, E.E. and Friedman, L.**, *Ion-Molecule Reactions*, Wiley-Interscience, New York, 1970.
21. **Franklin, J.L.**, Ed., *Ion-Molecule Reactions*, Plenum Press, New York, 1972.
22. **Ausloos, P.**, Ed., *Interactions Between Ions and Molecules*, Plenum Press, New York, 1975.
23. **Lias, S.G. and Ausloos, P.**, *Ion-Molecule Reactions. Their Role in Radiation Chemistry*, American Chemical Society, Washington, 1975.
24. **Ausloos, P.**, Ed., *Kinetics of Ion-Molecule Reactions*, Plenum Press, New York. 1979.
25. **Bowers, M.T.**, Ed., *Gas Phase Ion Chemistry*, Academic Press, New York, 1979.
26. **Almoster-Ferreira, M.A.**, Ed., *Ionic Processes in the Gas Phase*, Reidel, Dordrecht, 1984.
27. **Futrell, J.H.**, Ed., *Gaseous Ion Chemistry and Mass Spectrometry*, John Wiley & Sons, New York, 1986.
28. **Farrar, J.M. and Saunders, Jr., W.H.**, Eds., *Techniques for the Study of Ion-Molecule Reactions*, John Wiley & Sons, New York, 1988.
29. **Ausloos, P., Lias, S.G., and Dixon, D.**, Eds., *Structure, Reactivity and Thermochemistry of Ions*, Reidel, Dordrecht, 1987.

30. **Eyring, H, Hirschfelder, J.O., and Taylor, H.S.**, The theoretical treatment of chemical reactions produced by ionization processes. I. The ortho-para hydrogen conversion by alpha particles, *J. Chem. Phys.*, 4, 479, 1936.

31. **Vogt, E. and Wannier, G.H.**, Scattering of ions by polarization processes, *Phys. Rev.*, 95, 1190, 1954.

32. **Gioumousis, G. and Stevenson, D.P.**, Reactions of gaseous ions with gaseous molecules. II. Theory, *J. Chem. Phys.*, 29, 294, 1958.

33. **Langevin, P.M.**, Une formule fondamentale de theorie cinetique, *Ann. Chim. Phys.*, 5, 245, 1905.

34. **McDaniel, E.W.**, *Collision Phenomena in Ionized Gases*, John Wiley & Sons, New York, 1964.

35. **Henglein, A.**, Kinematics of ion-molecule reactions, in *Molecular Beams and Reaction Kinetics*, Schlier, C., Ed., Academic Press, New York, 1970.

36. **Su, T. and Bowers, M.T.**, Classical ion-molecule collision theory, in *Gas Phase Ion Chemistry*, Bowers, M.T., Ed., Academic Press, New York, 1979.

37. **Shirts, R.B.**, Collision theory and reaction dynamics, in *Gaseous Ion Chemistry and Mass Spectrometry*, Futrell, J.H., Ed., John Wiley & Sons, New York, 1986.

38. **Moran, T.F. and Hamill, W.H.**, Cross-sections of ion permanent dipole reactions by mass spectrometry, *J. Chem. Phys.*, 39, 1413, 1963.

39. **Theard, L.P. and Hamill, W.H.**, The energy dependence of cross-sections of some ion-molecule reactions, *J. Am. Chem. Soc.*, 84, 1134, 1962.

40. **Gupta, S.K., Jones, E.G., Harrison, A.G., and Myher, J.J.**, Reactions of thermal energy ions. VI. Hydrogen transfer reactions involving polar molecules, *Can. J. Chem.*, 45, 3107, 1967.

41. **Bowers, M.T. and Laudenschlager, J.B.**, Mechanism of charge transfer reactions of rare gas ions with the trans-, cis-, and 1,1-difluoroethylene geometric isomers, *J. Chem. Phys.*, 56, 4711, 1972.

42. **Su, T. and Bowers, M.T.**, Theory of ion-polar molecule collisions. Comparison with experimental charge transfer reactions of rare gas ions to geometric isomers of difluorobenzene and dichloroethylene, *J. Chem. Phys.*, 58, 3027, 1973.

43. **Su, T. and Bowers. M.T.**, Ion-polar molecule collisions: the effect of ion size on ion-polar molecule rate constants: the parameterization of the average dipole orientation theory, *Int. J. Mass Spectrom. Ion Phys.*, 12, 347, 1973.

44. **Bass, L., Su, T., and Bowers, M.T.**, Ion-polar molecule collisions. A modification of the average dipole orientation theory: the $\cos \theta$ model, *Chem. Phys. Lett.*, 34, 119, 1975.

45. **Su, T. and Bowers, M.T.**, Parameterization of the average dipole orientation theory: temperature dependence, *Int. J. Mass Spectrom. Ion Phys.*, 17, 211, 1975.

46. **Bohme, D.K.**, The kinetics and energetics of proton transfer, in *Interactions between Ions and Molecules*, Ausloos, P., Ed., Plenum Press, New York, 1975.

47. **Harrison, A.G., Lin, P.-H., and Tsang, C.W.**, Proton transfer reactions by trapped ion mass spectrometry, *Int. J. Mass Spectrom. Ion Phys.*, 19, 23, 1976.

48. **Huntress, W.T., Mosesman, M.M., and Elleman, D.D.**, Relative rates and their dependence on kinetic energy for ion-molecule reactions in ammonia, *J. Chem. Phys.*, 54, 843, 1971.

49. **Bowers, M.T., Su, T., and Anicich, V.G.**, Theory of ion-polar molecule collisions. Kinetic energy dependence of ion-polar molecule reactions: $CH_3OH^+ + CH_3OH \rightarrow CH_3OH_2^+ + CH_3O$, *J. Chem. Phys.*, 58, 5175, 1973.

50. **Su, T., Su, E.C.F., and Bowers, M.T.**, Ion-polar molecule collisions. Conservation of angular momentum in the average dipole orientation theory. The AADO theory, *J. Chem. Phys.*, 69, 2243, 1978.

51. **Su, T. and Bowers, M.T.**, Ion-polar molecular collisions: the average quadrupole orientation theory, *Int. J. Mass Spectrom. Ion Phys.*, 17, 309, 1975.

52. **Ikezoe, Y., Matsuoka, S., Takebe, M., and Viggiano, A.**, *Gas Phase Ion-Molecule Rate Constants Through 1986*, Maruzen Company, Ltd., Tokyo, 1987.

53. **Lindholm, E.**, Charge exchange and ion-molecule reactions observed in double mass spectrometers, in *Ion-Molecule Reactions in the Gas Phase*, Ausloos, P., Ed., American Chemical Society, Washington, 1966.

54. **Lindholm, E.**, Mass spectra and appearance potentials studied by use of charge exchange in a tandem mass spectrometer, in *Ion-Molecule Reactions*, Franklin, J.L., Ed., Plenum Press, New York, 1972.

55. **Harrison, A.G.**, Charge exchange mass spectrometry, in *Ionic Processes in the Gas Phase*, Almoster Ferreira, M.A., Ed., Reidel, Dordrecht, 1984.

56. **Bowers, M.T. and Elleman, D.D.**, Thermal energy charge transfer reactions of rare gas ions to methane, ethane, propane and silane. The importance of Frank-Condon factors, *Chem. Phys. Lett.*, 16, 486, 1972.

57. **Ausloos, P., Eyler, J.R., and Lias, S.G.**, Thermal energy charge transfer reactions involving CH_4 and SiH_4. Lack of evidence for non-spiralling collisions, *Chem. Phys. Lett.*, 30, 21, 1975.

58. **Li, Y.-H. and Harrison, A.G.**, Bimolecular reactions of trapped ions. XIII. Charge transfer from Ar^+, Kr^+ and N_2^+ to methane, *Int. J. Mass Spectrom. Ion Phys.*, 28, 289, 1978.

59. **Li, Y.-H. and Harrison, A.G.**, Unpublished results.

60. **Abramson, F.P. and Futrell, J.H.**, Ion-molecule reactions of methane, *J. Chem. Phys.*, 45, 1925, 1966.

61. **Huntress, W.T. and Pinnizotto, R.F.**, Product distribution and rate constants for ion-molecule reactions in water, hydrogen sulfide, ammonia and methane, *J. Chem. Phys.*, 59, 492, 1972.

62. **Hellner, L. and Sieck, L.W.**, High pressure photoionization mass spectrometry. Effect of internal energy and density on the ion-molecule reactions occurring in methyl, dimethyl and trimethyl amine, *Int. J. Chem. Kin.*, 5, 177, 1973.

63. **Herod, A.A., Harrison, A.G., and McAskill, N.A.**, Ion-molecule reactions in methyl fluoride and methyl chloride, *Can. J. Chem.*, 49, 2217, 1971.

64. **Solka, B.H. and Harrison, A.G.**, Bimolecular reactions of trapped ions. X. Reactions in methyl mercaptan and mixtures with methane, *Int. J. Mass Spectrom. Ion Phys.*, 14, 295, 1974.

65. **Harrison, A.G., Heslin, E.J., and Blair, A.S.**, Bimolecular reactions of trapped ions. IV. Ion-molecule reactions in ethane and mixtures with C_2H_2 and CD_4, *J. Am. Chem. Soc.*, 94, 2935, 1972.

66. **Solka, B.H., Lau, A.Y.K., and Harrison, A.G.**, Bimolecular reactions of trapped ions. VIII. Reactions in propane and propane-methane mixtures, *Can. J. Chem.*, 52, 1798, 1974.

67. **Solka, B.H. and Harrison, A.G.**, Bimolecular reactions of trapped ions. IX. Effect of method of preparation on rate of reaction of $C_2H_5^+$, *Int. J. Mass Spectrom. Ion Phys.*, 14, 125, 1974.

68. **Su, T. and Bowers, M.T.**, Ion-polar molecule collisions. The effect of molecular size on ion-polar molecule rate constants, *J. Am. Chem. Soc.*, 95, 7609, 1973.

69. **Su, T. and Bowers, M.T.**, Ion-polar molecule collisions. Proton transfer reactions of $C_4H_9^+$ ions with NH_3, CH_3NH_2, $(C_2H_5)_2NH$ and $(CH_3)_3N$, *J. Am. Chem. Soc.*, 95, 7611, 1973.

70. **Chesnavich, W.J., Su, T. and Bowers, M.T.**, Collisions in a non-central field: a variational and trajectory investigation of ion-dipole capture, *J. Chem. Phys.*, 72, 2641, 1980.

71. **Su, T. and Chesnavich, W.J.**, Parameterization of the ion-polar molecule collision rate constant by trajectory calculations, *J. Chem. Phys.*, 76, 5183, 1982.

72. **Hemsworth, R.S., Payzant, J.D., Schiff, H.I., and Bohme, D.K.**, Rate constants at 297 K for proton transfer reactions with NH_3. Comparison with classical theories and exothermicities, *Chem. Phys. Lett.*, 26, 4167, 1974.

73. **Mackay, G.I., Betkowski, L.D., Payzant, J.D., Schiff, H.I., and Bohme, D.K.**, Rate constants at 297 K for proton transfer reactions with HCN and CH_3CN. Comparisons with classical theories and exothermicity, *J. Chem. Phys.*, 80, 2919, 1976.

74. **Solka, B.H. and Harrison, A.G.**, Bimolecular reactions of trapped ions. XI. Rate and equilibria in proton transfer reactions of $CH_3SH_2^+$, *Int. J. Mass Spectrom. Ion Phys.*, 17, 379, 1975.

75. **Bohme, D.K., Mackay, G.I., and Schiff, H.I.**, Determination of proton affinities from the kinetics of proton transfer reactions. VII. The proton affinities of O_2, H_2, Kr, O, N_2, Xe, CO_2, CH_4, N_2O and CO, *J. Chem. Phys.*, 73, 4976, 1980.

76. **Field, F.H. and Lampe, F.W.**, Reactions of gaseous ions. VI. Hydride ion transfer reactions, *J. Am. Chem. Soc.*, 80, 5587, 1958.

77. **Ausloos, P. and Lias, S.G.**, Structure and reactivity of hydrocarbon ions, in *Ion-Molecule Reactions*, Franklin, J.L., Ed., Plenum Press, New York, 1972.

78. **Lias, S.G.**, Ion-molecule reactions in radiation chemistry, in *Interactions Between Ions and Molecules*, Ausloos, P., Ed., Plenum Press, New York, 1975.

79. **Searles, S.K. and Sieck, L.W.**, High pressure photoionization mass spectrometry. III. Reactions of $NO^+(X^1\Sigma^+)$ with C_3-C_7 hydrocarbons at thermal kinetic energies, *J. Chem. Phys.*, 53, 794, 1970.

80. **Lias, S.G., Eyler, J.R., and Auloos, P.**, Hydride ion transfer reactions involving saturated hydrocarbons and CCl_3^+, CCl_2H^+, CF_2Cl^+, CF_2H^+, CF_3^+, NO^+, $C_2H_5^+$, sec-$C_3H_7^+$ and t-$C_4H_9^+$, *Int. J. Mass Spectrom. Ion Phys.*, 19, 219, 1976.

81. **Ausloos, P. and Lias, S.G.**, Carbonium ions in radiation chemistry. Reactions of t-butyl ions with hydrocarbons, *J. Am. Chem. Soc.*, 92, 5037, 1970.

82. **Lias, S.G., Rebbert, R.E., and Ausloos, P.**, Gas phase radiolysis of hydrocarbon mixtures: determination of the charge recombination rate coefficient and absolute rate constants of the t-butyl ion through a competitive kinetic method, *J. Chem. Phys.*, 57, 2080, 1972.

83. **Hunt, D.F. and Harvey, J.M.**, Nitric oxide chemical ionization mass spectra of alkanes, *Anal. Chem.*, 47, 1965, 1975.

84. **Chai, R. and Harrison, A.G.**, Mixtures of NO with hydrogen and with methane as chemical ionization reagent gases, *Anal. Chem.*, 55, 969, 1983.

85. **Hunt, D.F. and Ryan, J.F.**, Chemical ionization mass spectrometry studies. Nitric oxide as a reagent gas. *J. Chem. Soc. Chem. Commun.*, 620, 1972.

86. **Doepker, R.D. and Ausloos, P.**, Gas-phase radiolysis of cyclopentane. Relative rates of H_2^--transfer reactions from various hydrocarbons to $C_3H_6^+$, *J. Chem. Phys.*, 44, 1951, 1965.

87. **Gordon, R., Doepker, R. and Auloos, P.**, Photoionization of propylene at 1236 Å. Reactions of $C_3D_6^+$ with added alkanes, *J. Chem. Phys.*, 44, 3733, 1965.

88. **Sieck, L.W. and Futrell, J.H.**, Reactions of $C_3H_6^+$ with C_3 and C_4 paraffins, *J. Chem. Phys.*, 45, 560, 1966.

89. **Lias, S.G. and Auloos, P.**, Structure and reactivity of $C_4H_8^+$ ions formed in the radiolysis of alkanes in the gas phase, *J. Am. Chem. Soc.*, 92, 1840, 1970.

90. **Sieck, L.W. and Searles, S.K.**, High pressure photoionization mass spectrometry. II. A study of thermal $H^-(H^0)$ and $H_2^-(H_2^0)$ transfer reactions occurring in alkane-olefin mixtures, *J. Am. Chem. Soc.*, 92, 2937, 1970.

91. **McMahon, T.B., Blint, R.J., Ridge, D.P., and Beauchamp, J.L.**, Determination of carbonium ion stabilities by ion cyclotron resonance spectroscopy, *J. Am. Chem. Soc.*, 94, 8934, 1972.

92. **Dawson, J.H.J., Henderson, W.G., O'Malley, R.M., and Jennings, K.R.**, Halogen ion transfer reactions in the gas phase ion chemistry of haloalkanes, *Int. J. Mass Spectrom. Ion Phys.*, 11, 61, 1973.

93. **Blint, R.J., McMahon, T.B., and Beauchamp, J.L.**, Gas-phase ion chemistry of fluoromethanes by ion cyclotron resonance spectroscopy. New techniques for the determination of carbonium ion stabilities, *J. Am. Chem. Soc.*, 96, 1269, 1974.

94. **Wieting, R.D., Staley, R.H., and Beauchamp, J.L.**, Relative stabilities of carbonium ions in the gas phase and solution. A comparison of cyclic and acyclic carbonium ions, acyl cations and cyclic halonium ions, *J. Am. Chem. Soc.*, 96, 7552, 1974.

95. **Sieck, L.W., Lias, S.G., Hellner, L. and Ausloos, P.**, Photoionization of C_4H_8 isomers. Unimolecular and bimolecular reactions of the $C_4H_8^+$ ion, *J. Res. Nat. Bur. Stand.*, 76A, 115, 1972.

96. **Abramson, F.P. and Futrell, J.H.**, Ionic reactions in unsaturated compounds. III. Propylene and the isomeric butenes, *J. Phys. Chem.*, 72, 1994, 1968.

97. **Myher, J.J. and Harrison, A.G.**, Ion-molecule reactions in propyne and allene, *J. Phys. Chem.*, 72, 1905, 1968.

98. **Bowers, M.T., Aue, D.H., and Elleman, D.D.**, Mechanisms of ion-molecule reactions in propene and cyclopropane, *J. Am. Chem. Soc.*, 94, 4255, 1972.

99. **Chin Lin, M.S. and Harrison, A.G.**, Ion-molecule reactions in isobutene, *Int. J. Mass Spectrom. Ion Phys.*, 17, 97, 1975.

100. **Herod, A.A. and Harrison, A.G.**, Effect of kinetic energy on ionic reactions in propylene and cyclopropane, *J. Phys. Chem.*, 73, 3189, 1969.

101. **Hughes, B.M. and Tiernan, T.O.**, Ionic reactions in gaseous cyclobutane, *J. Chem. Phys.*, 51, 4373, 1969.

102. **Sieck, L.W., Searles, S.K., and Ausloos, P.**, High pressure photoionization mass spectrometry. I. Unimolecular and bimolecular reactions of $C_4H_8^+$ from cyclobutane, *J. Am. Chem. Soc.*, 91, 7627, 1969.

103. **Henis, J.M.S.**, Ion-molecule reactions in olefins, *J. Chem. Phys.*, 52, 282, 1970.

104. **Gross, M.L. and Norbeck, P.**, Effects of vibrational energy on the rates of ion-molecule reactions, *J. Chem. Phys.*, 54, 3651, 1971.

105. **Sieck, L.W. and Ausloos, P.**, Photoionization of propylene, cyclopropane and ethylene. The effect of internal energy on the bimolecular reactions of $C_2H_4^+$ and $C_3H_6^+$, *J. Res. Nat. Bur. Stand.*, 76A, 253, 1972.

106. **Bowers, M.T., Elleman, D.D., O'Malley, R.M., and Jennings, K.R.**, Analysis of ion-molecule reactions in allene and propyne by ion cyclotron resonance, *J. Phys. Chem.*, 74, 2583, 1970.

107. **Ferrer-Correia, A.J. and Jennings, K.R.**, ICR mass spectra of fluoroalkenes. IV. Ion-molecule reactions in mixtures of ethylene and the fluoroethylenes, *Int. J. Mass Spectrom. Ion Phys.*, 11, 111, 1973.

108. **Ferrer-Correia, A.J., Jennings, K.R., and Sen Sharma, D.K.**, The use of ion-molecule reactions in the mass spectrometric location of double bonds, *Org. Mass Spectrom.*, 11, 867, 1976.

109. **Chai, R. and Harrison, A.G.**, Location of double bonds by chemical ionization mass spectrometry, *Anal. Chem.*, 53, 34, 1981.

110. **Meot-Ner (Mautner), M.**, Temperature and pressure effects in the kinetics of ion-molecule reactions, in *Gas Phase Ion Chemistry*, Bowers, M.T., Ed., Academic Press, New York, 1979.

111. **Kebarle, P.**, Higher order reactions, ion clusters and ion solvation, in *Ion-Molecule Reactions*, Franklin, J.L., Ed., Plenum Press, New York, 1972.

112. **Durden, D.A., Kebarle, P., and Good, A.**, Thermal ion-molecule reaction rate constants at pressures up to 10 torr with a pulsed mass spectrometer. Reactions in methane, krypton and oxygen, *J. Chem. Phys.*, 50, 805, 1969.

113. **Meot-Ner, M. and Field, F.H.**, Stability, association and dissociation in the cluster ions $H_3S^+nH_2S$, $H_3O^+nH_2O$ and $H_3S^+nH_2O$, *J. Am. Chem. Soc.*, 99, 998, 1977.

114. **Nielson, P.V., Bowers, M.T., Chan, M., Davidson, W.R., and Aue, D.H.**, Energy transfer in excited species. Rate and mechanism of dimerization of protonated amines and their neutral dimers, *J. Am. Chem. Soc.*, 100, 3649, 1978.

115. **Ferguson, E.E., Fehsenfeld, F.C., and Albritton, D.L.**, Ion chemistry of the earth's atmosphere, in *Gas Phase Ion Chemistry*, Bowers, M.T., Ed., Academic Press, New York, 1979.

116. **Christophoru, L.G.**, The lifetimes of metastable negative ions, *Adv. Electron. Electron Phys.*, 46, 55, 1978.

117. **Ho, A.G., Bowie, J.H., and Fry, A.**, Electron impact studies. LVII. Negative ion mass spectrometry of functional groups: simple esters, *J. Chem. Soc. B*, 530, 1971.

118. **Bowie, J.H.**, Electron impact studies. LXXXII. Negative ion mass spectrometry of functional groups. Collision induced spectra of the carboxyl group, *Org. Mass Spectrom.*, 9, 304, 1974.

119. **Bowie, J.H. and Janposri, S.**, Electron impact studies. LX. Negative ion mass spectrometry of functional groups. Skeletal rearrangements in arylnitro compounds, *Org. Mass Spectrom.*, 5, 945, 1971.

120. **Bowie, J.H. and Janposri, S.**, Electron impact studies. CI. Negative ion mass spectra of the carbonyl group, the aryl-CH_2-CO- and aryl-$(CH_2)_2$-CO- systems, *Org. Mass Spectrom.*, 10, 1117, 1975.

121. **Christophoru, L.G. and Stockdale, J.A.D.**, Dissociative electron attachment to molecules, *J. Chem. Phys.*, 48, 1956, 1968.

122. **Viggiano, A.A. and Paulson, J.F.**, Temperature dependence of associative detachment reactions, *J. Chem. Phys.*, 79, 2241, 1983.

123. **Lieder, C.A. and Brauman, J.I.**, A technique for detection of neutral products in gas-phase ion-molecule reactions, *Int. J. Mass Spectrom. Ion Phys.*, 16, 307, 1975.

124. **Lieder, C.A. and Brauman, J.I.**, Detection of neutral products in gas-phase ion-molecule reactions, *J. Am. Chem. Soc.*, 96, 4028, 1974.

125. **DePuy, C.H., Gronert, S., Mullin, A., and Bierbaum, V.M.**, Gas-phase S_N2 and E2 reactions of alkyl halides, *J. Am. Chem. Soc.*, 112, 8650, 1990.

126. **Olmstead, W.N. and Brauman, J.I.**, Gas-phase nucleophilic displacement reactions, *J. Am. Chem. Soc.*, 99, 4219, 1977.

127. **Brauman, J.I.**, Factors influencing thermal ion-molecule rate constants, in *Kinetics of Ion-Molecule Reactions*, Ausloos, P., Ed., Plenum Press, New York, 1979.

128. **Jones, M.E. and Ellison, G.B.**, A gas-phase E2 elimination reaction: methoxide ion and bromopropane, *J. Am. Chem. Soc.*, 111, 1645, 1989.

129. **Nibbering, N.M.M.**, Gas-phase reactions of organic anions, *Adv. Phys. Org. Chem.*, 24, 1, 1988.

130. **Bartmess, J.E. and McIver, Jr., R.T.**, The gas-phase acidity scale, in *Gas Phase Ion Chemistry*, Bowers, M.T., Ed., Academic Press, New York, 1979.

131. **Bohme, D.K.**, Ion chemistry: A new perspective, *Trans. R. Soc. Can.*, Ser. IV, XIX, 265, 1981.

132. **Farneth, W.E. and Brauman, J.I.**, Dynamics of proton transfer involving delocalized negative ions in the gas phase, *J. Am. Chem. Soc.*, 98, 7891, 1976.

133. **Ferguson, E.E., Fehsenfeld, F.C., and Schmeltekopf, A.L.**, Ion molecule reaction rates measured in a discharge afterglow, in *Chemical Reactions in Electrical Discharges*, Adv. Chem. Ser. 80, American Chemical Society, Washington, 1969.

134. **Fahey, D.W., Bohringer, H, Fehsenfeld, F.C., and Ferguson, E.E.**, Reaction rate constants for $O_2^-(H_2O)_n$, n = 0 to 4, with O_3, NO, SO_2 and CO_2, *J. Chem. Phys.*, 76, 1799, 1982.

135. **Babcock, L.M. and Streit, G.E.**, Negative ion molecule reactions of SF_4, *J. Chem. Phys.*, 75, 3864, 1981.

136. **Kebarle, P. and Chowdhury, S.**, Electron affinities and electron-transfer reactions, *Chem. Rev.*, 87, 513, 1987.

137. **Fehsenfeld, F.C. and Ferguson, E.E.**, Laboratory studies of negative ion reactions with atmospheric trace constituents, *J. Chem. Phys.*, 61, 3181, 1974.

138. **Payzant, J.D., Cunningham, A.J., and Kebarle, P.**, Kinetics and rate constants of reactions leading to hydration of NO_2^- and NO_3^- in gaseous oxygen, argon and helium containing traces of water, *Can. J. Chem.*, 50, 2230, 1972.

139. **McDonald, R.N. and Chowdhury, A.K.**, Nucleophilic reactions of F_3C^- at sp^2 and sp^3 carbon in the gas phase. Characterization of carbonyl addition products, *J. Am. Chem. Soc.*, 105, 7267, 1983.

140. **Beaty, E.C., Branscomb, L.M., and Patterson, P.L.**, Mobilities of oxygen negative ions in oxygen, *Bull. Am. Phys. Soc.*, 9, 535, 1964.

141. **Brønsted, J.N.**, Einize bemurkungen uber den begriff der sauren und basen, *Recl. Trav. Chim. Pay-Bas*, 42, 718, 1923.

142. **Beauchamp, J.L. and Buttrill, S.E.**, Proton affinities of H_2S and H_2O, *J. Chem. Phys.*, 48, 1783, 1968.

143. **Long, J. and Munson, B.**, Proton affinities of some oxygenated compounds, *J. Am. Chem. Soc.*, 95, 2427, 1973.

144. **Aue, D.H., Webb, H.M., and Bowers, M.T.**, Quantitative proton affinities, ionization potentials and hydrogen affinities of alkyl amines, *J. Am. Chem. Soc.*, 98, 311, 1976.

145. **Kebarle, P.**, Pulsed electron high pressure mass spectrometer, in *Techniques for the Study of Ion-Molecule Reactions*, Farrar, J.M. and Saunders, Jr., W.H., Eds., John Wiley & Sons, New York, 1988.

146. **Schiff, H.I. and Bohme, D.K.**, Flowing afterglow studies at York University, *Int. J. Mass Spectrom. Ion Phys.*, 16, 167, 1975.

147. **Mason, R.S., Fernandez, M.T., and Jennings, K.R.**, Thermodynamics of some proton transfer reactions, *J. Chem. Soc. Faraday Trans. 2*, 83, 89, 1987.

148. **McLuckey, S.A., Cameron, D., and Cooks, R.G.**, Proton affinities from dissociation of proton-bound dimers, *J. Am. Chem. Soc.*, 103, 1313, 1981.

149. **Isa, K., Omote, T., and Amaya, M.**, New rules concerning the formation of protonated amino acids from protonated peptides using the proton affinity order determined from collisionally activated decomposition spectra, *Org. Mass Spectrom.*, 25, 620, 1990.

150. **Taft, R.W.**, Gas-phase proton transfer equilibria, in *Proton Transfer Reactions*, Caldin, E.F. and Gold, V., Eds., Chapman and Hall, London, 1975.

151. **Aue, D.H. and Bowers, M.T.**, Stabilities of positive ions from equilibrium gas-phase basicity measurements, in *Gas Phase Ion Chemistry*, Bowers, M.T., Ed., Academic Press, New York, 1979.

152. **Lias, S.G., Liebman, J.F., and Levin, R.D.**, Evaluated gas phase basicities and proton affinities of molecules; heats of formation of protonated molecules, *J. Phys. Chem. Ref. Data*, 13, 695, 1984.

153. **Bartmess, J.E.**, Gas-phase equilibrium affinity scales and chemical ionization mass spectrometry, *Mass Spectrom. Rev.*, 8, 297, 1989.

154. **Lias, S.G., Bartmess, J.E., Liebman, J.F., Holmes, J.L., Levin, R.D., and Miallard, W.G.**, Gas-phase ion and neutral thermochemistry, *J. Phys. Chem. Ref. Data*, 17 (Suppl. 1), 1988.

155. **Carroll, T.X., Smith, S.R., and Thomas, T.D.**, Correlations between proton affinity and core electron ionization potentials for double-bonded oxygen. Site of protonation of esters, *J. Am. Chem. Soc.*, 97, 659, 1975.

156. **Mills, B.E., Martin, R.L., and Shirley, D.A.**, Further studies of core binding energy-proton affinity correlations in molecules, *J. Am. Chem. Soc.*, 98, 2380, 1976.

157. **Benoit, F.M. and Harrison, A.G.**, Predictive value of proton affinity-ionization energy correlations involving oxygenated molecules, *J. Am. Chem. Soc.*, 99, 3980, 1977.

158. **Brown, R.S. and Tse, A.**, Determination of circumstances under which the correlation of core binding energy and gas-phase basicity or proton affinity breaks down, *J. Am. Chem. Soc.*, 102, 5222, 1980.

159. **Lau, Y.K. and Kebarle, P.**, Substituent effects on the intrinsic basicity of benzene: proton affinities of substituted benzenes, *J. Am. Chem. Soc.*, 98, 7452, 1976.

160. **Liauw, W.G. and Harrison, A.G.**, Site of protonation in the reaction of gaseous Brønsted acids with halobenzene derivatives, *Org. Mass Spectrom.*, 16, 388, 1981.

161. **Mason, R.S., Milton, D., and Harris, F.**, Proton transfer to the fluorine atom in fluorobenzene: a dramatic temperature dependence in the gas phase, *J. Chem. Soc. Chem. Commun.*, 1453, 1987.

162. **Tsang, C.W. and Harrison, A.G.**, The chemical ionization of amino acids, *J. Am. Chem. Soc.*, 98, 1301, 1976.

163. **Meot-Ner (Mautner), M. and Sieck, L.W.**, Proton affinity ladders from variable-temperature equilibrium measurements. I. A reevaluation of the upper proton affinity range, *J. Am. Chem. Soc.*, 113, 4448, 1991.

164. **Solomon, J.J. and Field, F.H.**, Stability of some C_7 tertiary alkyl carbonium ions, *J. Am. Chem. Soc.*, 98, 1025, 1976.

165. **Solomon, J.J. and Field, F.H.,** Reversible reactions of gaseous ions. X. The intrinsic stability of the norbornyl ion, *J. Am. Chem. Soc.,* 98, 1567, 1976.
166. **Cox, J.D. and Pilcher, G.,** *Thermochemistry of Organic and Organometallic Compounds,* Academic Press, New York, 1970.
167. **Stull, D.F. and Prophet, H.,** *JANAF Thermochemical Tables,* National Standards Ref. Data Ser., Nat. Bur. Stand., Washington, D.C., 1971.
168. **Hotop, H. and Lineburger, W.C.,** Binding energies in atomic negative ions II, *J. Phys. Chem. Ref. Data,* 14, 731, 1985.
169. **Drzaic, P.S., Marks, J., and Brauman, J.I.,** Electron photodetachment from gas phase molecular anions, in *Gas Phase Ion Chemistry, Vol. 3 Ions and Light,* Bowers, M.T., Ed., Academic Press, Orlando, 1984.
170. **Mead, R.D., Stevens, A., and Lineburger, W.C.,** Photodetachment in negative ion beams, in *Gas Phase Ion Chemistry, Vol. 3 Ions and Light,* Bowers, M.T., Ed., Academic Press, Orlando, 1984.
171. **Tiernan, T.O.,** Reactions of negative ions, in *Interactions Between Ions and Molecules,* Ausloos, P., Ed., Plenum Press, New York, 1975.
172. **Christophoru, L.G. and Goans, R.E.,** Low energy (<1 eV) electron attachment to molecules in very high pressure gases: C_6H_6, *J. Chem. Phys.,* 60, 4244, 1974.
173. **Janousek, B.K. and Brauman, J.I.,** Electron affinities, in *Gas Phase Ion Chemistry,* Bowers, M.T., Ed., Academic Press, New York, 1979.
174. **Ervin, K.M., Gronert, S., Barlow, S.E., Gilles, M.K., Harrison, A.G., Bierbaum, V.M., DePuy, C.H., Lineberger, W.C., and Ellison, G.B.,** Bond strengths of ethylene and acetylene, *J. Am. Chem. Soc.,* 112, 5750, 1990.
175. **Boand, G., Houriet, R., and Gäumann, T.,** Gas-phase acidity of aliphatic alcohols, *J. Am. Chem. Soc.,* 105, 2203, 1983.
176. **Graul, S.T., Schnute, M.E., and Squires, R.R.,** Gas-phase acidities of carboxylic acids and alcohols from collision-induced dissociation of dimer cluster ions, *Int. J. Mass Spectrom. Ion Processes,* 96, 181, 1990.
177. **Bartmess, J.E., Scott, J.A., and McIver, Jr., R.T.,** Substituent and solvation effects in gas-phase acidity scales, *J. Am. Chem. Soc.,* 101, 6056, 1979.

Chapter 3

INSTRUMENTATION FOR CHEMICAL IONIZATION

I. INTRODUCTION

In CIMS, ionization of the sample of interest is accomplished by interaction of the sample molecules with gaseous reagent ions, viz.,

$$R^{\pm} + S \rightarrow S^{\pm} + R \tag{1}$$

The yield of sample ions, $S^{\pm}$, depends on the magnitude of the ion/molecule reaction rate constant, on the concentration (or partial pressures) of the reacting species, $R^{\pm}$ and S, and on the time available for reaction between the two. As discussed in the previous chapter, the reaction rate constant frequently is close to the limiting value determined by the ion/molecule collision rate as established by ion-dipole and ion-induced dipole interactions. The partial pressures of reacting species required and the reaction time available are interrelated quantities which depend on the instrument used. These factors and other factors peculiar to CIMS are discussed in this chapter; no attempt is made to cover in any detail the general instrumentation of mass spectrometry since this aspect has been covered adequately elsewhere.[1-4]

II. MEDIUM PRESSURE MASS SPECTROMETRY

The majority of chemical ionization studies have been carried out using sources only slightly modified from those used in electron ionization studies; indeed, most are capable of operating in either the electron or chemical ionization mode. In such sources, diffusional loss of ions to the walls and removal of ions from the source by the necessary extraction fields limits the source residence time to no more than ~10 μsec. To achieve adequate sensitivity in the chemical ionization mode, it is necessary to operate with reagent gas pressures in the range 0.3 to 1.0 torr and to use sample pressures of the order of 10^{-3} to 10^{-4} torr in the ion source. At the same time, the mass analysis region must be maintained in the region of 10^{-6} torr to prevent excessive ion scattering and the pressure in the ion source housing must be maintained below approximately 10^{-4} torr to prevent damage to the electron emitting filament and to prevent ion/molecule collisions between the source ion exit slit and the point of entry into the mass analysis system. In addition, for magnetic deflection instruments, the pressure in the source housing must be kept at this level to prevent discharge to ground of the high ion accelerating potential that normally is used.

To achieve these conditions the ion source must be made as gas-tight as possible, the pumping speed at the source housing must be as high as practicable, and the analyzer region must be differentially pumped. The ion source is carefully constructed to ensure a gas-tight assembly and the aperture for entry of the ionizing electron beam is reduced, relative to that used in electron ionization sources, typically to 0.1 to 0.2 mm^2. The ion exit slit is reduced both in width and length, typical widths being of the order of 5×10^{-2} mm and total area 0.1 to 0.2 mm^2; a somewhat larger slit area can be accommodated if a thicker plate is used to limit the conductance of the slit. Connections for introduction of the reagent gas and the sample, the latter by heated inlet, direct insertion probe, or gas chromatograph, should be made as gas-tight as possible. Most manufacturers now offer a combined chemical ionization/electron ionization source as a standard feature and, therefore, have incorporated at least adequate pumps on their instruments; in the early days of chemical ionization, a number of workers reported details of the modification of existing electron ionization mass spectrometers for operation in the chemical ionization mode.[5–10] Included in the modifications usually has been a redesign of the source housing and pumping leads to provide adequate source region pumping speeds; the simple addition of a high speed diffusion pump is not adequate if the pumping line leading to the source is of small diameter, as was frequently the case in older instruments.

With source pressures of 0.5 to 1.0 torr, electrons with 70 eV energy will have a very short range and it is usual to provide the capability of employing electron energies up to 400 eV so that the electrons can penetrate further into the source. However, even under these conditions, the electron beam is likely to be completely attenuated before it reaches the position where the electron trap normally is installed in an electron ionization source. Consequently, the filament current cannot be regulated to give a constant trap current, as is done under electron ionization conditions, but rather is regulated to give a constant electron current to the source block. It is hoped that, if a constant flux of electrons impinges on the source block, a constant flux will pass through the electron entrance aperture and enter the ionization chamber. Electron impact ionization of the reagent gas normally is employed and, with most commonly used reagent gases, the increased pressure does not lead to an unacceptably short filament lifetime. However, if oxidizing reagent gases, such as nitric oxide or oxygen are used, filament lifetimes are severely reduced. For these cases, a Townsend discharge ion source[11] and a glow discharge ion source[12] have been constructed to replace the electron ionization source. An additional advantage of discharge ion sources is that the source can be operated at much lower temperatures than are possible when a heated filament is used.

A problem associated with the operation of magnetic sector instruments under chemical ionization conditions is that of electrical discharges through the gas (at 0.3 to 1.0 torr) between the ion source, which is held at a high potential (positive or negative), and a ground located somewhere in an inlet line. This discharge may occur in the line admitting the reagent gas, in the line admitting volatile samples, or in the line leading to a pressure measuring device. The discharge occurs when the electric field causes a breakdown of the gas resulting in a flow of electrons toward the anode (the ion source in positive ion CI) and of positive ions towards the cathode (ground in positive ion CI). The breakdown voltage is a function of the electric field strength and the gas density and varies considerably with the nature of the gas; breakdown is particularly prevalent, however, at the pressures used in chemical ionization. The discharge is quenched at higher pressures (above ~50 torr), and this fact has been used in several inlet systems where a large pressure drop is created through an electrically nonconducting capillary leak. The disadvantage of such an approach is that large sample backing pressures are required, creating a problem when small sample sizes or compounds of low volatility are used; in addition, it is impossible to measure the source pressure directly through such a lead. Indeed, most commercial mass spectrometers equipped for chemical ionization make no provision for measuring the source pressure and one must either estimate the source pressure indirectly or develop other criteria for deciding that satisfactory CI conditions have been established. An approach to estimating the reagent gas pressure from ion/molecule reaction rates has been published[13] for the cases where methane or isobutane are the reagent gases. For the same gases one of the xylenes has been used as a test compound to indicate good CI conditions. In CH_4 CI ratios of m/z 107:m/z 106 $\geq$ 10:1 and m/z 106:m/z 91 $\geq$ 10:1 are taken as indicative of satisfactory CI conditions. For i-C_4H_{10} CI ratios of m/z 107:m/z 106 $\geq$ 3:1 and m/z 106:m/z 91 $\geq$ 10:1 are used as indicators of satisfactory CI conditions.

Where it is desired to avoid capillary lines or to attach a pressure measuring device to the source of a magnetic sector instrument, several approaches to solving the discharge problem have been published.[6,14] As an alternative, the entire sample inlet system and pressure measuring device can be floated at source potential and powered via a mains isolation transformer while being located behind a protective screen.[15] An interesting point is that pyrex glass becomes a good conductor of electricity at 250 to 300°C, and it may be necessary to insert a short length of tubing made of quartz or other insulating glass in the lines. These discharge problems are avoided when quadrupole mass spectrometers are used since much lower potentials are applied to the ion source and focusing electrodes.

When a magnetic deflection instrument is operated under negative ion chemical ionization conditions, the negative ions can be readily detected if the potential of the ion source is appreciably more negative with respect to earth than the potential of the first dynode of the electron multiplier. For example, a negative ion accelerated from an ion source at –8 kV will still have 5 keV kinetic energy when it collides with the first dynode of the multiplier held at –3 kV and secondary electrons will be produced. However, if the two voltages are similar, or if a quadrupole mass filter is used, where the ions have energies of typically 10 to 25 eV, no ion signal will be measured unless the detection system is modified. If the first dynode of the multiplier is held at ground potential or at a positive potential and the last dynode is held at +2 to +6 kV, secondary electrons will be produced in good yield. However, the output signal must be conducted by a suitable, high voltage, floating coaxial signal feedthrough to a preamplifier and appropriate circuitry to reference the signal to ground potential.[16,17] While this is possible, the approach has the associated problem that stray electrons in the vacuum system are detected efficiently

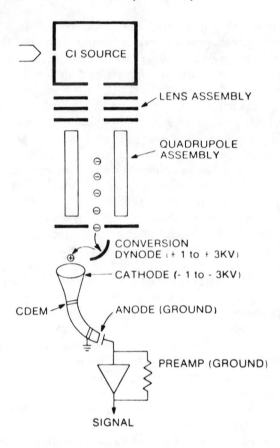

FIGURE 1.   Negative ion continuous dynode electron multiplier incorporating conversion dynode. (From Stafford, G.C., *Environ. Health Perspect.*, 36, 85, 1980. With permission.)

causing an increase in the background signal which may be as much as a factor of 30.[17] An alternative approach, which is more commonly used, is to convert the negative ions to positive ions by impact on a conversion dynode biased at a suitably high positive potential. The positive ions so produced are then accelerated to the first, negatively biased, dynode of the multiplier and are detected in the usual way. A typical arrangement, due to Stafford and co-workers[17,18] is shown in Figure 1. It is believed that the major mechanisms responsible for the negative to positive ion conversion are sputtering, fragmentation and charge stripping, or inversion, of the negative ion. Normally, the gain of an electron multiplier on impact of positive ions decreases with increasing ion mass; by contrast, the gain for negative ions using the conversion dynode multiplier system increases with increasing ion mass.[17,18]

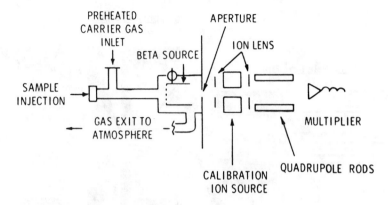

FIGURE 2.   Low-volume atmospheric pressure ionization mass spectrometer. (From Horning, E.C., Horning, M.G., Carroll, D.I., Dzidic, I., and Stillwell, R.N., *Anal. Chem.*, 45, 936, 1973. With permission.)

III. ATMOSPHERIC PRESSURE CHEMICAL IONIZATION (APCI)

In APCI, primary positive and negative ions are generated in a flowing carrier gas at atmospheric pressure either by β-particles from a ^{63}Ni source[19,20] or by a corona discharge.[21,22] The terminal ions present in the flowing gas are sampled through a small aperture and focused by a suitable lens arrangement into a quadrupole mass filter for analysis. Samples are introduced into the flowing gas stream by injection in a solvent, by direct insertion on a probe, or as an effluent from a gas or liquid chromatograph; if ambient air is used as the carrier gas, trace impurities in the air can be detected directly. A considerable amount of gas passes through the aperture and to maintain the mass analysis system at a suitable pressure ($<10^{-5}$ torr) requires efficient pumping.

The development of APCI sources has proceeded along two lines. One design[19,23,24] is similar to an electron capture detector in that there is a low-volume chamber (~1 cm^3) coupled to the analysis system through a small (typically 25 μ) orifice. This design normally uses a pure gas at a flow rate of 10 to 100 ml min^{-1} as the carrier gas. Because of the small size and the low flow rate the source is always heated to minimize absorption on the walls of the source. Analysis is best performed by introducing one component at a time into the source; consequently, it is particularly suited for coupling to a gas chromatograph or liquid chromatograph to act as an ultrasensitive detector. A typical low-volume APCI instrument is shown schematically in Figure 2.

In the second approach,[22,25,26] illustrated in Figure 3, the ionization source (usually a corona discharge) is suspended in a large volume source chamber with the reaction volume being defined by the volume between the corona point and the sampling orifice, which is usually larger (~100 μ). The larger orifice can be accommodated because the cryogenic pumping provides a very

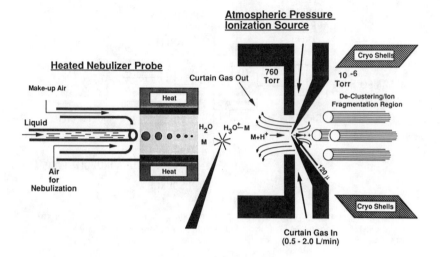

FIGURE 3.   Large-volume atmospheric pressure ionization source, illustrated with liquid nebulizer sample introduction. (Copyright 1990 Sciex. All rights reserved. Reproduced with permission.)

high pumping speed in the analysis region. The sampling orifice is separated from the reaction region by a stream of high purity dry nitrogen which serves not only to prevent particulate matter from clogging the orifice but also serves to assist in declustering the solvated analyte ions often formed. A high flow rate (up to 9 L s^{-1}) of carrier or sample gas (including ambient air) is used. Since the ions and sample molecules react in an essentially wall-less region, which is continuously swept by the carrier gas to exclude molecules that do desorb from the walls, the source may be operated at ambient temperatures.

Theoretical considerations[23,27] show that the terminal ions formed in the weak plasma produced using a ^{63}Ni source often are dominated by ions characteristic of trace impurities rather than by the major component(s). In effect, concentration and reaction time parameters are such that the ionization ends up residing on the most stable positive or negative ion species. Hence, the desirability of introducing one component at a time. The use of a corona discharge gives a much higher ionization rate as well as a shorter ion source residence time, due to the high electric field between the corona point and the orifice, with the result[26] that the response to one species is not influenced to a great extent by the presence of others at trace levels. Consequently, the large volume source has seen extensive application in direct mixture analysis, particularly ambient air. A major advantage of APCI is that detection limits at the parts per trillion level can be attained, although, as will be discussed in Chapter 4, some analytes show very low sensitivity in APCI.

In a rigorously cleaned APCI instrument using extremely high purity nitrogen gas, the major primary charged species are N_3^+ and N_4^+ as positive ions and thermal energy electrons as negative species. In practical operation, at least trace amounts of water will be present leading to appreciable intensities for $H^+(H_2O)_n$ in the positive ion regime. In addition, trace amounts of oxygen are likely to be present leading to formation of O_2^- (and possibly $O_2^-[H_2O]_n$) as well as thermal energy electrons in the negative ion mode.[28] The solvated ions will be the dominant ions when ambient air is used as the carrier gas. The mechanisms of formation of the reagent ions observed in APCI will be discussed in Chapter 4. With ambient air as the carrier gas, and frequently with other carrier gases, the analyte ions will be solvated by H_2O, leading to a loss of sensitivity because the analyte signal may be spread over several species. In the instrument illustrated in Figure 3, some declustering occurs in the dry nitrogen curtain gas; declustering also is carried out by collision-induced dissociation in the lens region following the sampling orifice.[26,29] Since the bond energies in the cluster species are lower than the bond energies in the core ion, declustering without fragmentation of the core analyte ion is possible by adjustment of the lens voltage, i.e., the collision energy.

The subject of mass spectrometry with ion sources operating at atmospheric pressure has been reviewed recently.[30]

IV. FOURIER TRANSFORM MASS SPECTROMETRY (FTMS)

Although the majority of chemical ionization studies have been carried out in medium pressure sources or in APCI instruments, chemical ionization could be achieved at much lower pressure if the reaction time could be made correspondingly longer by trapping the reactant and product ions in appropriate devices. One such device is the trapped ion cyclotron resonance (ICR) spectrometer[31,32] in which the ions are trapped and constrained to a circular orbit by crossed electric and magnetic fields. Primary ions are produced by a short pulse of electrons and allowed to react with the analyte during the trapping time which can be of the order of seconds. Compared to the 10^{-5} s reaction time available in medium pressure CI it is apparent that partial pressures can be reduced by a factor of 10^5 to 10^6 and still achieve the same extent of conversion in bimolecular ion/molecule reactions in the ICR experiment. Obviously, if the primary ions are mass analyzed immediately after the ionizing, pulse electron ionization mass spectra will be obtained; the device therefore has the advantage that EI or CI mass spectra can be obtained merely by changing the pulse sequence and timing.

Early studies of the analytical uses of the trapped ICR method showed[33-35] that the technique suffered from severe limitations. Mass resolution was limited to about one amu at m/z 200 while the mass range was less than about m/z 280. In addition, the scan rate was very low. While these limitations do not seriously compromise the use of the technique for fundamental studies of the kinetics and equilibria of gas phase ion/molecule reactions,[32] they do compromise the analytical uses. The introduction of Fourier transform techniques to ICR by Comisarow and Marshall[36] has resolved many of these problems. In the trapped ion cyclotron resonance cell the ions move in circular orbits with a cyclotron frequency ω_c (in hertz) given by[37]

$$\omega_c = 1.537 \times 10^7 \, zB/m \qquad (2)$$

where z is the unit charge of the ion, B is the magnetic field strength in tesla, and m is the ion mass in atomic mass units. If a radio frequency (r.f.) field having the same frequency as the cyclotron frequency of the ion is applied to the cell, the ion absorbs energy causing its orbital radius and velocity to increase but without a change in cyclotron frequency. In addition, after only a few rotations, the ions all are moving together coherently. This coherent ion package will induce an oscillating image current on parallel receiving plates. Digitization and fast Fourier transformation of this image current yields a frequency domain spectrum which can be converted into a mass spectrum. A complete mass spectrum is obtained by applying an r.f. pulse covering a wide frequency range for a short period of time (typically 1 msec). The image current then contains frequencies for all the m/z ratios present and the Fourier transform yields the frequency spectrum and, thus, the complete mass spectrum.

The Fourier transform approach solves many of the problems inherent in the ICR method. Clearly, a mass spectrum can be obtained rapidly over a mass range which is limited only by the frequency range of the excitation pulse. The mass resolution increases with the length of time the transient decay signal is sampled. The length of time this decay signal can be sampled is determined primarily by collisional damping wherein ion/neutral collisions cause the ion packet to lose its coherent motion. Low resolution spectra are obtained with total pressures $>10^{-5}$ torr in the cell while high resolution spectra are obtained at pressures below 10^{-8} torr in the cell.

Chemical ionization mass spectra are obtained by FTMS using, typically, 10^{-6} torr pressure of reagent gas with sample pressures of the order of 10^{-8} torr, with a time delay between the ionizing event and ion detection of the order of seconds. Clearly, under these conditions, relatively low resolution mass spectra will be obtained. Higher resolution can be achieved by pulsed-valve addition of the reagent gas providing reasonable pressures for ion formation and low pressures during the detection period.[38,39] An alternative approach involves the use of a dual-cell FTMS[40] where the reagent ions are produced in one cell at higher pressures and transferred to the second cell for reaction with the analyte and for product ion detection at lower pressures. More recently, external ion sources have been coupled to Fourier transform instruments[41–43] allowing the generation of reagent ions at high pressures and introduction of selected reagent ions into the FTMS cell for reaction at lower pressures. In the absence of such sophisticated devices, "self-chemical ionization", in which ions produced from the sample by the initial EI process react with the sample by ion/molecule reactions, may be used.[44] The FTMS method also is ideally suited for the study of the reactions of metal ions, produced by laser desorption from a metal target, with organic substrates;[45–49] this method has been called "reagentless CI" by Freiser.[37]

Two reviews[50,51] of FTMS have appeared recently, the first emphasizing the developments in chemical applications, the second emphasizing the advances in instrumentation and techniques.

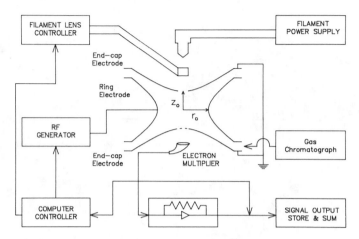

FIGURE 4. Schematic diagram of the quadrupole ion trap mass spectrometer. (From *American Laboratory*, 15, 54, 1983. With permission.)

V. QUADRUPOLE ION TRAP MASS SPECTROMETER

The quadrupole ion trap is another device in which ions can be stored and the reaction time varied, thus providing the opportunity to carry out chemical ionization at low pressures. The trap, shown schematically in Figure 4, consists of three cylindrically symmetrical electrodes, two endcaps, and a ring, each of which has accurately machined hyperbolic internal surfaces. Ions produced by a pulse of electrons, or injected from an external source, are trapped in the device by application of suitable DC voltages (including grounding) to the endcap electrodes and radiofrequency (r.f.) voltage to the ring electrode. This trap, first proposed by Paul and Steinwedel,[52] has seen extensive use for physical and chemical studies of gas-phase ions[53] but saw little application for analytical studies because of limited resolution and sensitivity.

The current, rapidly increasing analytical use of the ion trap results from two developments reported in 1984 by Stafford et al.[54] First, it was found that a relatively high pressure ($\sim 10^{-3}$ torr) of a light gas such as He in the ion trap dampened the motion of the ions in the trap and led to substantial increases in both the resolution and the sensitivity. Second, the technique of mass-selective ion ejection was developed. As the amplitude of the r.f. voltage applied to the ring electrode is increased the ions develop unstable trajectories along the axis of symmetry, and in order of increasing m/z value, exit the device through holes in one of the endcaps and are detected by an electron multiplier detector. The theory of operation has been discussed in detail elsewhere.[53,55,56] As marketed commercially, the ion trap mass spectrometer has an upper mass/charge limit of 650 Da e^{-1}, although there is very active research to increase both the mass range and the resolution.[56,57]

The attraction that the ion trap mass spectrometer shares with the FTMS is that either electron ionization or chemical ionization mass spectra can be

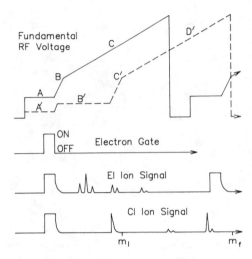

FIGURE 5.   Operation of the ion trap in electron ionization and chemical ionization mode. Sequence of operation discussed in text. (Copyright 1985 Finnigan Corporation. All rights reserved. Reproduced with permission.)

obtained simply by changing the timing and sequence of voltages applied to the various elements of the device. The method of operation of the ion trap to obtain EI and CI mass spectra can be understood with reference to Figure 5. In the EI mode of operation (A,B,C,) the magnitude of the r.f. voltage indicated by A permits the storage of sample ions of interest during the ionization period. The r.f. voltage is then increased rapidly, B, to the level appropriate for the selected low m/z at which the mass spectral scan is to commence. The ramp C represents the slow scan of the r.f. voltage during which the analyte ions are mass-selectively ejected and a mass spectrum is obtained. For chemical ionization operation the sequence A′,B′,C′,D′ is followed. Initially the r.f amplitude is set at A′ to permit formation and storage of the primary ions of the reagent gas and to allow them to react with the reagent gas. During this time, the r.f. amplitude is sufficiently low so that high mass ions, which might be formed by electron ionization of the analyte, are not efficiently stored. The r.f. amplitude is then increased to B′ to permit storage of the sample ions produced by reaction of the reagent ions with the sample. Finally, during C′, the r.f. amplitude is ramped to the value corresponding to the low mass limit of the mass spectral scan which is acquired during the ramp D′.

A number of studies have examined the use of the quadrupole ion trap mass spectrometer to acquire both positive ion[58-60] and negative ion[61] chemical ionization mass spectra. The spectra obtained are in general agreement with those obtained by medium pressure mass spectrometry except that cluster ions are generally absent at the low pressures (typically 1×10^{-3} torr, primarily He bath gas), and more fragmentation is observed because of the higher mean

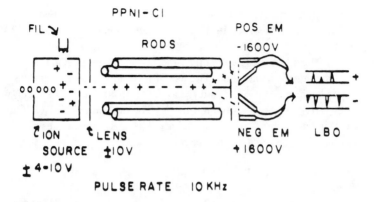

FIGURE 6.   Pulsed positive/negative ion chemical ionization (PPNICI) mass spectrometer. (FIL — filament, EM — electron multiplier, LBO — light beam oscillograph) (From Hunt, D.F., Stafford, G.C., Crow, F.W., and Russell, J.W., *Anal. Chem.*, 48, 2098, 1976. With permission.)

energies of the ions in the quadrupole field. It has been estimated[62] that the average energy of the ions in the ion trap corresponds to an equivalent temperature of ~600 to 700 K. An advantage which the ion trap mass spectrometer shares with FTMS is that by application of suitable r.f. and DC voltages to the trap, individual ions in the reagent ion plasma can be selected[63,64] for chemical ionization reaction with the substrate. At the same, time it might be noted that electrons are not efficiently trapped in the quadrupole field and electron capture chemical ionization sensitivities should be quite low.

The developments in and applications of the quadrupole ion trap mass spectrometer have been reviewed recently.[55,56,65]

VI. PULSED POSITIVE ION/NEGATIVE ION CHEMICAL IONIZATION

Over the past 15 years, negative ion chemical ionization has become a powerful analytical method. In many cases it offers increased sensitivity and selectivity and, frequently, it provides information which is complementary to positive ion chemical ionization. There is an obvious advantage in obtaining both the positive ion and the negative ion chemical ionization mass spectra of a sample. The quadrupole mass filter transmits negative ions and positive ions with equal facility and Hunt et al.[66] have described a pulsed mode of operation which permits simultaneous recording of positive and negative chemical ionization mass spectra. Their arrangement is shown in Figure 6. By pulsing both the ion source potential (± 4 to 10 V) and the focusing lens potential (± 10 to 30 V) at a rate of 10 kHz, packets of positive and negative ion are detected by separate continuous dynode electron multipliers operating at suitable potentials, with, in the original design, the two signals being recorded simultaneously on separate channels of a light beam oscillograph. Alternatively, the two signals can be fed to a data system for processing.

By suitable choice of reagent gas, various types of chemical ionization mass spectra may be recorded in the positive and negative ion modes.[66] With nitrogen as the reagent gas, $N_2^{+\cdot}$, reacting by charge exchange, is the dominant positive ion reagent species, while quasi-thermal electrons provide the ionizing reagent in the negative ion mode. The use of methane as the reagent gas gives the Brønsted acid reagents CH_5^+ and $C_2H_5^+$ in the positive ion mode and quasi-thermal electrons in the negative ion mode. The addition of CH_3ONO to the methane reagent results in the Brønsted base, CH_3O^-, in the negative ion mode without changing the positive ion reagents.

VII. SAMPLE INTRODUCTION IN CIMS

The major requirement of any sample introduction system in CIMS is that the sample be introduced as a vapor, with a partial pressure of the order of 10^{-3} torr, into the ion source containing the reagent gas plasma. Samples may be introduced either from a heated inlet system or from a direct insertion probe; one must ensure that these modes of sample introduction do not provide a route for discharge of the high voltage to ground when a sector instrument is used. When samples are introduced as a vapor from a heated inlet, the pressure differential between the sample and the reagent gas may not be sufficient to permit significant flow through the flow-regulating leak. In this case, it is necessary to add reagent gas to the sample reservoir to create the necessary pressure differential.

The combination of a gas chromatograph coupled to a mass spectrometer is one of the most powerful and sensitive analytical methods available. Consequently, close to 2000 papers per year are published in the gas chromatography/mass spectrometry (GC/MS) field,[67] many employing chemical ionization as the ionization method of choice. The early history of GC/MS has been covered by McFadden;[68] more recent developments can be found in most standard texts and will not be reviewed here. Most modern GC/MS instruments are equipped with capillary columns and fast pumping with the result that the total effluent from the chromatograph can be accommodated by the mass spectrometer with the reagent gas being added separately to effect the chemical ionization method of choice.

For mixtures which are not sufficiently volatile for gas chromatography, separation by liquid chromatography (LC) or supercritical fluid chromatography (SFC) often is possible and there is considerable interest and activity in interfacing these chromatographic techniques to a mass spectrometer.[69–74] For both, the problems are greater than for GC/MS because of the much larger amount of gas which must be handled by the mass spectrometer pumping system, although this problem is considerably alleviated by the use of microbore columns since they operate at much lower flow rates. One of the earliest LC/MS interfaces[75] involved direct injection of part of the eluent into the ion source with the solvent serving as the reagent gas for chemical ionization;

because of its simplicity this method still sees considerable use. APCI instruments also have seen considerable use in LC/MS studies,[76–78] although solvation of the analyte ions by the polar solvents used may be a problem. These methods allow only chemical ionization spectra to be obtained; either EI or CI mass spectra of the analyte may be obtained using the moving belt interface[79] or the monodisperse aerosol generation interface.[80] The SFC/MS combination is more recent and less well developed. Brønsted acid[81,82] and negative ion[83] chemical ionization have been demonstrated using SFC/MS. CO_2 is a common working fluid for SFC and the use of CO_2 as a moderating gas for electron capture chemical ionization (ECCI) in SFC/MS also has been investigated briefly;[84] in the positive ion mode CO_2 will act as a charge exchange reagent with CO_2^+ as the dominant ion. Despite the advances made, it is likely that for the large fragile molecules amenable to LC or SFC, ionization methods such as flow-fast atom bombardment, electrospray, or ionspray will play a more important role than chemical ionization.

The requirement that the sample be in the gas phase for analysis by chemical ionization (or electron ionization) places severe limitations on the method for involatile, thermally labile molecules. If evaporated from a probe, the energy required to overcome the bonding between molecules and the bonding of molecules to the surface of the sample holder may exceed the energy required to break chemical bonds within the molecule and extensive sample decomposition may result. Some success in vaporizing intact labile peptides has been achieved by rapid heating (up to $12°C\ s^{-1}$) of samples dispersed on an inert Teflon surface.[85] The dispersion on the Teflon surface enhances volatility by decreasing both intermolecular forces and molecule/surface forces. If the heating is sufficiently rapid, the kinetic energy of the molecules is increased rapidly and these forces are overcome before appreciable energy becomes concentrated in intramolecular modes and results in bond rupture. Rather similar results have been obtained by Cotter[86] and by Amster and co-workers[87,88] using a pulsed laser to vaporize the sample from a probe with ionization by chemical ionization.

An alternative approach which has been named variously as direct chemical ionization, in-beam ionization, desorption chemical ionization, or surface ionization[89] involves dispersion of the sample on a suitable probe (often heatable) which then is inserted directly into the reagent ion plasma in the ion source. In our experience the use of such a direct exposure probe is beneficial for both CI and EI of involatile and labile molecules. In chemical ionization, the method was first employed by Baldwin and McLafferty in 1973.[90] Several designs of such direct exposure probes have been reported;[91–97] the most commonly used consists of a loop of electrically resistant wire (such as platinum) which is inserted into the reagent plasma or electron beam. With a programmed heating current, the sample is "distilled" from the probe and ionized without contacting other surfaces. However, at the upper temperatures of the probe, significant decomposition of the sample may occur in some cases.[98]

VIII. INSTRUMENTATION FOR COLLISION-INDUCED DISSOCIATION STUDIES

By suitable choice of reagent gas, it frequently is possible by chemical ionization to confine the ionization of the substrate largely, if not completely, to a single ion. While this is useful for molecular mass determination, fragment ions usually are necessary to provide structural information. For pure substrates this fragmentation often can be achieved by using more energetic reagent ions, i.e., making the ionization reaction more exothermic. This aspect will be discussed in Chapter 5. However, when it is desired to identify the individual components of a complex mixture without prior separation, this simple approach fails because the spectrum becomes too complex to interpret.

The desire to carry out such identifications in complex mixtures has led to the field of tandem mass spectrometry or mass spectrometry/mass spectrometry (MS/MS).[99,100] In this technique, the components of the mixture are ionized, usually by chemical ionization, to produce the simplest possible collection of ions representative of the neutrals present in the mixture. In the case where one is attempting to identify an individual component, the ion thought to represent that component is selected by a first mass spectrometer and subjected to collision-induced dissociation (CID) by collision with a suitable gas in a gas cell, with the products of dissociation being identified by mass analysis in a second mass spectrometer. Identification is made either by interpretation of the CID mass spectrum from first principles or by comparison with the CID mass spectrum of a pure authentic sample. Clearly, each precursor can be selected in turn and the component, in principle, identified. In effect, the selection of the reactant ion for CID serves the same purpose as separation of the components in time by chromatographic methods, while the identification from the CID mass spectrum is equivalent to identification of the separated components from their electron or chemical ionization mass spectra. Hence the abbreviation MS/MS is used to contrast the method with GC/MS. A technique which frequently is used in detection of a known compound in a complex matrix, even when chromatographic separation is used, is selected reaction monitoring. In this technique, the first mass spectrometer is set to transmit the known precursor ion while the second mass spectrometer is set to transmit a known fragment ion or stepped between two or more known fragment ions. This approach not only provides better limits of detection, by decreasing the chemical noise, but also increases the confidence that a positive identification has been made.[101] Clearly, the CID method can be used, in principle, to provide information on the identity of a pure component which gives only one ion in the CI mass spectrum.

Although the instrumentation and techniques of tandem mass spectrometry have been reviewed in detail recently,[100] a brief review of the major aspects will be presented here. Collision-induced dissociation studies have been carried out extensively using double-focusing mass spectrometers with the configuration B/E (Figure 7a) where the magnetic sector precedes the electrostatic sector. With this instrument, ionization of the sample in the source

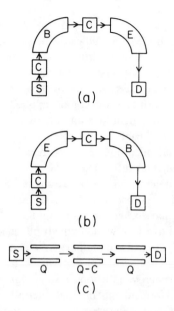

FIGURE 7. Schematic of analyzer arrangements for collision-induced dissociation studies.

is followed by mass selection of the ion of interest by the magnetic sector B, with collision-induced dissociation occurring in the collision cell between the two analyzers. The ionic products of dissociation (M_p) will have a kinetic energy

$$T_p = \left(M_p / M_i \right) T_i$$

where T_i and M_i are the kinetic energy and mass of the precursor ion. The second stage electrostatic sector is a kinetic energy analyzer and, by scanning it, the masses of the product ions can be established from their kinetic energies relative to the precursor kinetic energy. This method represents the so-called MIKES (mass-analyzed ion kinetic energy spectroscopy) technique. Similar results also can be obtained, using either the B/E instrument or the E/B instrument (Figure 7b) by using so-called linked scan methods. If the magnetic (B) and electric (E) fields are scanned down at a constant B/E ratio (that required to transmit the precursor ion), the ionic products of dissociation of the precursor ion in the collision cell between the source and the first sector will be successively detected and mass identified either from the magnetic field or the electric field required to transmit them. This linked scan provides a higher resolution spectrum of the product ions than the MIKES spectrum, where the peaks may overlap due to kinetic energy release in the fragmentation process. However, the B/E linked scan is more difficult experimentally and requires close tracking of the magnetic and electric fields, usually under computer control.

More recently, triple quadrupole mass spectrometers (Figure 7c) have been developed for such MS/MS studies. In these instruments, the precursor ion is mass selected by the first quadrupole, is collisionally dissociated in the second quadrupole which is pressurized by a collision gas, and the ionic products of dissociation are mass analyzed by the final quadrupole. The second quadrupole is operated with only r.f. voltage applied to it and, thus, serves to focus the precursor and fragment ions into the final quadrupole. In particular, one might note that several triple quadrupole mass spectrometers interfaced to atmospheric pressure chemical ionization sources have been reported.[102–104]

The major fundamental difference between CID studies carried out on triple quadrupole instruments and sector instruments lies in the kinetic energies of the ions subjected to collision with the neutral gas. In the instruments depicted in Figure 7a and 7b, the ions entering the collision cell have kinetic energies in the range 4 to 8 keV while in the triple quadrupole instrument, the ions entering the collision quadrupole have kinetic energies in the range 5 to 100 eV. In the collision process, some of the kinetic energy of the ion is converted into internal energy of the ion causing fragmentation. In many cases, the mean energy transformed into internal energy in low and high energy collisions is similar, although the distribution of internal energies extends to higher values for the high energy collisions.[105]

Low energy collision-induced experiments also can be carried out on Fourier transform instruments and quadrupole ion trap instruments.[100] In both cases, a r.f. pulse is used to eliminate all ions of mass lower than the one of interest from the cell, following which a r.f. pulse of appropriate frequency is applied to excite the ion of interest to a higher kinetic energy. During a wait period, these excited ions collide with an appropriate collision gas and suffer dissociation, the fragment ions are trapped and are mass analyzed when a mass analyzing pulse of the appropriate form is applied. It might also be noted that hybrid instruments incorporating magnetic and electric sectors with quadrupole stages also have been use for CID studies at both low and high collision energies.[100]

Two further types of CID studies deserve comment. It frequently is the case that compounds which are closely related structurally will show a common ion in their CID mass spectra and one may wish to identify all the precursors which give this common fragment. This can be accomplished readily using the triple quadrupole instrument by setting the final quadrupole to transmit the common fragment while scanning the initial quadrupole to identify the mass of the precursor which leads to this fragment on collision-induced dissociation. Similar results can be obtained on double-focusing instruments (Figure 7a or 7b) by a linked scan in which the magnetic and electric fields are scanned, while keeping B^2/E constant, to detect the precursors giving the product ion in collision-induced dissociations in the cell prior to the first analyzer.

Finally, it sometimes is possible to identify classes of compounds from the mass of the neutral fragment lost on collision-induced dissociation, e.g., carboxylate ions $RCOO^-$ formed in negative ion CI of acids tend to lose CO_2 (44 amu) on CID. Thus, the identification of all ions which lose 44 amu on CID serves to identify carboxylic acids. Such a scan is simple in the triple quadrupole instrument where the first and third quadrupoles can be scanned simultaneously but offset by the mass of the neutral lost. A similar scan in B/E or E/B instruments requires a complex linked scan.[100]

Some results from collision-induced dissociation studies of ions formed by chemical ionization are discussed in Chapter 6.

REFERENCES

1. **Roboz, J.**, *Introduction to Mass Spectrometry: Techniques and Applications*, Interscience, New York, 1968.
2. **Rose, M.E. and Johnstone, R.A.W.**, *Mass Spectrometry for Chemists and Biochemists*, Cambridge University Press, London, 1982.
3. **Watson, J.T.**, *Introduction to Mass Spectrometry*, 2nd Ed., Raven Press, New York, 1985.
4. **White, F.A. and Wood, G.M.**, *Mass Spectrometry: Applications to Science and Engineering*, John Wiley & Sons, New York, 1986.
5. **Michnowicz, J. and Munson, B.**, Studies in chemical ionization mass spectrometry, *Org. Mass Spectrom.*, 4, 481, 1970.
6. **Futrell, J.H. and Wojcik, L.H.**, Modification of a high resolution mass spectrometer for chemical ionization studies, *Rev. Sci. Instrum.*, 42, 244, 1971.
7. **Beggs, D., Vestal, M.L., Fales, H.M. and Milne, G.W.A.**, A chemical ionization mass spectrometer source, *Rev. Sci. Instrum.*, 42, 1578, 1971.
8. **Hogg, A.M.**, Modification of an AEI MS-12 mass spectrometer for chemical ionization, *Anal. Chem.*, 44, 227, 1972.
9. **Yinon, J. and Boettger, H.G.**, Modification of an AEI/GEC MS-9 high-resolution mass spectrometer for electron impact/ chemical ionization studies, *Chem. Instrum.*, 4, 103, 1972.
10. **Garland, W.A., Weinkam, R.J., and Trager, W.F.**, Relatively simple modification of an AEI MS-902 high-resolution mass spectrometer to chemical ionization studies, *Chem. Instrum.*, 5, 271, 1974.
11. **Hunt, D.F., McEwen, C.N., and Harvey, T.M.**, Positive and negative chemical ionization mass spectrometry using a Townsend discharge ion source, *Anal. Chem.*, 47, 1730, 1975.
12. **Schneider, B., Breuer, M., Hartmann, H., and Budzikiewicz, H.**, Chemical ionization with aggressive gases — a simple glow discharge source for sector field instruments, *Org. Mass Spectrom.*, 24, 216, 1989.
13. **Hancock, R.A., Walder, R., and Weigel, H.**, Chemical ionization mass spectrometry: a method of estimating reagent gas pressure in the ion source based on kinetic data, *Org. Mass Spectrom.*, 14, 507, 1979.
14. **Illes, A.J., Bowers, M.T., and Meisels, G.G.**, Sample introduction and pressure measuring system for chemical ionization mass spectrometers, *Anal. Chem.*, 53, 1551, 1981.
15. **Jennings, K.R.**, Chemical ionization mass spectrometry, in *Gas Phase Ion Chemistry*, Bowers, M.T., Ed., Academic Press, New York, 1979.
16. **Smit, A.L.C. and Field, F.H.**, Gaseous anion chemistry. Formation and reactions of OH^-. Reactions of anions with N_2O. OH^- negative ion chemical ionization mass spectrometry, *J. Am. Chem. Soc.*, 99, 6471, 1977.

17. **Stafford, G.C.**, Instrumental aspects of positive and negative ion chemical ionization mass spectrometry, *Environ. Health Perspect.*, 36, 85, 1980.

18. **Stafford, G., Reeher, J., Smith, R., and Story, M.**, A novel negative ion detection system for a quadrupole mass spectrometer, in *Dynamic Mass Spectrometry*, Vol. 5, Price, D. and Todd, J.F.J., Eds., Heyden and Son, London, 1978.

19. **Horning, E.C., Horning, M.G., Carroll, D.I., Dzidic, I., and Stillwell, R.N.**, New picogram detection system based on a mass spectrometer with an external ion source at atmospheric pressure. *Anal. Chem.*, 45, 936, 1973.

20. **Carroll, D.I., Dzidic, I., Stillwell, R.N., Horning, M.G., and Horning, E.C.**, Subpicogram detection system for gas phase analysis based upon atmospheric pressure ionization (API) mass spectrometry, *Anal. Chem.*, 46, 706, 1974.

21. **Dzidic, I., Carroll, D.I., Stillwell, R.N., and Horning, E.C.**, Comparison of ions formed in nickel-63 and corona discharges using nitrogen, argon, isobutane, ammonia, and nitric oxide as reagents in atmospheric pressure ionization mass spectrometry, *Anal. Chem.*, 48, 1763, 1976.

22. **Lane, D.A., Thomson, B.A., Lovett, A.M., and Reid, N.M.**, Real-time tracking of industrial emissions through populated areas using a mobile APCI mass spectrometer system, *Adv. Mass Spectrom.*, 8, 1480, 1980.

23. **Siegel, N.W. and McKeown, M.C.**, Ions and electrons in the electron capture detector. Quantitative detection by atmospheric pressure ionization mass spectrometry, *J. Chromatogr.*, 122, 397, 1976.

24. **Grimsrud, E.P., Kim, S.H., and Gobby, P.L.**, Measurement of ions with a pulsed electron capture detector by mass spectrometry, *Anal. Chem.*, 51, 223, 1979.

25. **French, J.B. and Reid, N.M.**, Real-time targeted compound monitoring in air using the TAGA™ 3000 atmospheric pressure chemical ionization mass spectrometer, in *Dynamic Mass Spectrometry*, Vol. 6, Price, D. and Todd, J.F.J., Eds., Heyden and Son, London, 1981

26. **French, J.B., Thomson, B.A., Davidson, W.R., Reid, N.M., and Buckley, J.A.**, Atmospheric pressure chemical ionization mass spectrometry, in *Mass Spectrometry in Environmental Sciences*, Karasek, F.W., Hutzinger, O., and Safe, S., Eds., Plenum Press, New York, 1985.

27. **Siegel, M.W. and Fite, W.L.**, Terminal ions in weak atmospheric plasmas. Applications of atmospheric pressure ionization to trace impurity analysis in gases, *J. Phys. Chem.*, 80, 2871, 1976.

28. **Dzidic, I., Carroll, D.I., Stillwell, R.N., Horning, M.G., and Horning, E.C.**, Studies of negative ions by atmospheric pressure ionization mass spectrometry, *Adv. Mass Spectrom.*, 7, 319, 1978.

29. **Kambara, H. and Kanomata, I.**, Determination of impurities in gases by atmospheric pressure ionization mass spectrometry, *Anal. Chem.*, 49, 270, 1977.

30. **Bruins, A.P.**, Mass spectrometry with ion sources operating at atmospheric pressure, *Mass Spectrom. Rev.*, 10, 53, 1991.

31. **Lehman, T.A. and Bursey, M.M.**, *Ion Cyclotron Resonance Spectrometry*, John Wiley & Sons, New York, 1976.

32. **Kemper, P.R. and Bowers, M.T.**, Ion cyclotron resonance spectrometry, in *Techniques for the Study of Ion-Molecule Reactions*, Farrar, J.M. and Saunders, Jr., W.H., Eds., John Wiley & Sons, New York, 1988.

33. **Henis, J.M.S.**, An ion cyclotron resonance study of the ion-molecule reactions in methanol, *J. Am. Chem. Soc.*, 90, 844, 1968.

34. **Henis, J.M.S.**, Analytical implications of ion cyclotron resonance spectrometry, *Anal. Chem.*, 41(10), 22A, 1969.

35. **Gross, M.L. and Wilkins, C.L.**, Ion cyclotron resonance spectrometry: recent advances of analytical interest, *Anal. Chem.*, 43(14), 65A, 1971.

36. **Comisarow, M.B. and Marshall, A.G.**, Fourier transform ion cyclotron resonance spectroscopy, *Chem. Phys. Lett.*, 25, 282, 1974.

37. **Freiser, B.S.**, Fourier transform mass spectrometry, in *Techniques for the Study of Ion-Molecule Reactions*, Farrar, J.M. and Saunders, Jr., W.H., Eds., John Wiley & Sons, New York, 1988.

38. **Carlin, T.S. and Freiser, B.S.**, Pulsed valve addition of collision and reagent gases in Fourier transform mass spectrometry, *Anal. Chem.*, 55, 571, 1983.

39. **Laude, Jr., D.A., Johlman, C.L., Brown, R.S., Ijames, C.F., and Wilkins, C.L.**, Pulsed-valve chemical ionization for gas chromatography Fourier transform mass spectrometry, *Anal. Chim. Acta*, 178, 67, 1985.

40. **Cody, R.B., Kinsinger, J.A., Ghaderi, S., Amster, I.J., McLafferty, F.W., and Brown, C.G.**, Developments in analytical Fourier transform mass spectrometry, *Anal. Chim. Acta*, 178, 43, 1985.

41. **McIver, Jr., R.T., Hunter, R.L., and Bowers, W.D.**, Coupling a quadrupole mass spectrometer and a Fourier transform mass spectrometer, *Int. J. Mass Spectrom. Ion Processes*, 64, 67, 1985.

42. **Kofel, P, Allemann, M., Kellerhals, Hp., and Wanczek, K.P.**, External generation of ions in ICR spectroscopy, *Int. J. Mass Spectrom. Ion Processes*, 65, 97, 1985.

43. **Kofel, P., and McMahon, T.B.**, A high pressure external ion source for Fourier transform ion cyclotron resonance spectrometry, *Int. J. Mass Spectrom. Ion Processes*, 98, 1, 1990.

44. **Ghaderi, S., Kulkarni, P.S., Ledford, E.B., Wilkins, C.L., and Gross, M.L.**, Chemical ionization Fourier transform mass spectrometry, *Anal. Chem.*, 53, 428, 1981.

45. **Burnier, R.C., Byrd, G.D., and Freiser, B.S.**, Copper(I) chemical ionization— mass spectrometric analysis of esters and ketones, *Anal. Chem.*, 52, 1641, 1980.

46. **Burnier, R.C., Byrd, G.D. and Freiser, B.S.**, Gas-phase reaction of Fe^+ with ketones and ethers, *J. Am. Chem. Soc.*, 104, 3565, 1982.

47. **Byrd, G.D., Burnier, R.C., and Freiser, B.S.**, Gas-phase ion-molecule reactions of Fe^+ and Ti^+ with alkanes, *J. Am. Chem. Soc.*, 104, 5944, 1982.

48. **Freiser, B.S.**, Applications of laser ionization/Fourier transform mass spctrometry to the study of metal ions and their clusters in the gas phase, *Anal. Chim. Acta*, 178, 137, 1985.

49. **Chou, C.-C. and Long, S.R.**, Chemical ionization Fourier transform mass spectrometry of chemical warfare simulants using laser-produced metal ions, *Appl. Optics*, 29, 4981, 1990

50. **Wilkins, C.L., Chowdhury, A.K., Nuwaysir, L.M., and Coates, M.L.**, Fourier transform mass spectrometry. Current status, *Mass Spectrom. Rev.*, 8, 67, 1989.

51. **Marshall, A.G. and Grosshans, P.B.**, Fourier transform ion cyclotron resonance mass spectrometry: the teenage years, *Anal. Chem.*, 63, 215A, 1991.

52. **Paul, W and Steinwedel, H.**, Apparatus for separating charged particles of different specific charges, U.S. Patent 2,939,952, 7 June 1960.

53. **March, R.E. and Hughes, R.J.**, *Quadrupole Storage Mass Spectrometry*, John Wiley & Sons, New York, 1989.

54. **Stafford, Jr., G.C., Kelley, P.E., Syka, J.E.P., Reynolds, W.E., and Todd, J.F.J.**, Recent improvements in and analytical applications of advanced ion trap technology, *Int. J. Mass Spectrom. Ion Phys.*, 60, 85, 1984.

55. **Todd, J.F.J.**, Ion trap mass spectrometry-past, present, future (?), *Mass Spectrom. Rev.*, 10, 3, 1991.

56. **Todd, J.F.J. and Penman, A.D.**, The recent evolution of the quadrupole ion trap mass spectrometer— an overview, *Int. J. Mass Spectrom. Ion Processes*, 106, 1, 1991.

57. **Kaiser, Jr., R.E., Cooks, R.G., Stafford, Jr., G.C., Syka, J.E.P., and Hemberger, P.H.**, Operation of a quadrupole ion trap mass spectrometer to achieve high mass/charge ratios, *Int. J. Mass Spectrom. Ion Processes*, 106, 79, 1991.

58. **Brodbelt, J.S., Louris, J.N., and Cooks, R.G.**, Chemical ionization in an ion trap mass spectrometer, *Anal. Chem.*, 59, 1278, 1987.

59. **Boswell, S.M., Mather, R.E., and Todd, J.F.J.**, Chemical ionization in the ion trap: a comparative study, *Int. J. Mass Spectrom. Ion Processes*, 99, 139, 1990.

60. **Dorey, R.S.**, Increased fragmentation in methane chemical ionization in the ion trap detector, *Org. Mass Spectrom.*, 24, 973, 1989.

61. **McLuckey, S.A., Glish, G.L., and Kelley, P.E.**, Collision-activated dissociation of negative ions in an ion trap mass spectrometer, *Anal. Chem.*, 59, 1670, 1987.

62. **Nourse, B.D. and Kenttämaa, H.I.**, Effective ion temperatures in a quadrupole ion trap, *J. Phys. Chem.*, 94, 5809, 1990.

63. **Berberich, D.W., Hail, M.E., Johnson, J.V., and Yost, R.A.**, Mass selection of reagent ions for chemical ionization in quadrupole ion trap and triple quadrupole mass spectrometers, *Int. J. Mass Spectrom. Ion Processes*, 94, 115, 1989.

64. **Strife, R.J. and Keller, P.R.**, Ion trap chemical ionization. RF/DC for isolating unique reactant ions, *Org. Mass Spectrom.*, 24, 201, 1989.

65. **Nourse, B.D. and Cooks, R.G.**, Aspects of recent developments in ion trap mass spectrometry, *Anal. Chim. Acta*, 228, 1, 1990.

66. **Hunt, D.F., Stafford, F.C., Crow, F.W., and Russell, J.W.**, Pulsed positive negative ion chemical ionization mass spectrometry, *Anal. Chem.*, 48, 2098, 1976.

67. **Evershed, R.P.**, Analysis of mixtures by mass spectrometry Part I. Developments in gas chromatography/mass spectrometry, in *Mass Spectrometry, Specialist Periodical Report*, Vol. 10, Rose, M.E., Ed., Royal Society of Chemistry, Cambridge, 1989.

68. **McFadden, W.H.**, *Technique of Combined Gas Chromatography/Mass Spectrometry: Applications in Organic Analysis*, Wiley-Interscience, New York, 1973.

69. **Covey, T.R., Lee, E.D., Bruins, A.P., and Henion, J.D.**, Liquid chromatography/mass spectrometry, *Anal. Chem.*, 58, 1451A, 1986.

70. **Games, D.E.**, Applications of high performance liquid chromatography/ mass spectrometry (LC/MS) in food chemistry, in *Applications of Mass Spectrometry in Food Science*, Gilbert, J., Ed., Elsevier Applied Science, London, 1987.

71. **Catlow, D.A. and Rose, M.E.**, Analysis of mixtures by mass spectrometry. II. Techniques other than gas chromatography/mass spectrometry, in *Mass Spectrometry, Specialist Periodical Report*, Vol. 10, Rose, M.E., Ed., Royal Society of Chemistry, Cambridge, 1989.

72. **Arpino, P.**, Combined liquid chromatography mass spectrometry. Part I. Coupling by means of a moving belt interface, *Mass Spectrom. Rev.*, 8, 35, 1989.

73. **Smith, R.D., Kalinoski, H.T., and Udseth, H.R.**, Fundamentals and practice of supercritical fluid chromatography—mass spectrometry, *Mass Spectrom. Rev.*, 6, 445, 1987.

74. **Smith, R.D., Wright, B.W., and Yonker, C.R.**, Supercritical fluid chromatography: current status and prognosis, *Anal. Chem.*, 60, 1323A, 1988.

75. **Arpino, P.J., Baldwin, M.A., and McLafferty, F.W.**, Liquid chromatography—mass spectrometry. II. Continuous monitoring, *Biomed. Mass Spectrom.*, 1, 80, 1974.

76. **Carroll, D.I., Dzidic, I., Stillwell, R.N., Haegele, K.D., and Horning, E.C.**, Atmospheric pressure ionization mass spectrometry: corona discharge ion source for use in liquid chromatography—mass spectrometry—computer studies, *Anal. Chem.*, 47, 2369, 1975.

77. **Kambara, H.**, Sample introduction system for atmospheric pressure ionization of nonvolatile compounds, *Anal. Chem.*, 54, 143, 1982.

78. **Shushan, B., Fulford, J.E., Thomson, B.A., Davidson, W.R., Danylewych, L.M., Ngo, A., Nacson, S., and Tanner, S.D.**, Recent applications of triple quadrupole mass spectrometry to trace chemical analysis, *Int. J. Mass Spectrom. Ion Phys.*, 46, 225, 1983.

79. **McFadden, W.H., Schwartz, H.L., and Evans, S.**, Direct analysis of liquid chromatographic effluents, *J. Chromatogr.*, 122, 389, 1976.

80. **Willoughby, R.C. and Browner, R.F.**, Monodisperse aerosol generation interface for combining liquid chromatography with mass spectrometry, *Anal. Chem.*, 56, 2626, 1984.

81. **Kalinoski, H.T., Wright, B.W., and Smith, R.D.**, Ammonia and methane chemical ionization mass spectra of acid and carbamate pesticides using direct supercritical fluid injection, *Biomed. Environ. Mass Spectrom.*, 13, 33, 1986.

82. **Kalinoski, H.T., Wright, B.W., and Smith, R.D.**, Chemical ionization mass spectra of high molecular weight biologically active compounds following supercritical fluid chromatography, *Biomed. Environ. Mass Spectrom.*, 18, 64, 1989.

83. **Roach, J.A.F., Sphon, J.A., Easterling, J.A., and Colvey, E.M.**, Capillary supercritical fluid chromatography/negative ion chemical ionization of trichothecenes, *Biomed. Environ. Mass Spectrom.*, 18, 64, 1989.

84. **Sim, P.G. and Elson, C.M.**, Ion chemistry of PAHs under negative ion CI with CO_2, *Rapid Commun. Mass Spectrom.*, 2, 137, 1988.

85. **Beuhler, R.J., Flanigan, E., Greene, L.J., and Friedman, L.**, Proton transfer mass spectrometry of peptides. A rapid heating technique for underivatized peptides containing arginine, *J. Am. Chem. Soc.*, 96, 3990, 1974.

86. **Cotter, R.J.**, Laser desorption chemical ionization mass spectrometry, *Anal. Chem.*, 52, 1767, 1980.

87. **Amster, I.J., Land, D.P., Hemminger, J.C., and McIver, Jr., R.T.**, Chemical ionization of laser-desorbed neutrals in a Fourier transform mass spectrometer, *Anal. Chem.*, 61, 184, 1989.

88. **Speir, J.P., Borman, G.S., Cornett, D.S., and Amster, I.J.**, Controlling the dissociation of peptide ions using laser desorption/chemical ionization Fourier transform mass spectrometry, *Anal. Chem.*, 63, 65, 1991.

89. **Cotter, R.J.**, Mass spectrometry of non-volatile compounds: desorption from extended probes, *Anal. Chem.*, 52, 1589A, 1980.

90. **Baldwin, M.A. and McLafferty, F.W.**, Direct chemical ionization of relatively involatile samples. Application to underivatized oligopeptides, *Org. Mass Spectrom.*, 7, 1353, 1973.

91. **Hunt, D.F., Shabanowitz, J., Botz, F.K., and Brent, D.A.**, Chemical ionization mass spectrometry of salts and thermally labile organics with field desorption emitters as solids probes, *Anal. Chem.*, 49, 1161, 1977.

92. **Hansen, G. and Munson, B.**, Surface chemical ionization mass spectrometry, *Anal. Chem.*, 50, 1130, 1978.

93. **Cotter, R.J.**, Probe for direct exposure of solid samples to the reagent gas in a chemical ionization mass spectrometer, *Anal. Chem.*, 51, 317, 1979.

94. **Carroll, D.I., Dzidic, I., Horning, M.G., Montgomery, F.E., Nowlin, J.G., Stillwell, R.N., Thenot, J.–P., and Horning, E.C.**, Chemical ionization mass spectrometry of non-volatile organic compounds, *Anal. Chem.*, 51, 1858, 1979.

95. **Hansen, G. and Munson, B.**, Chemical ionization mass spectrometry of thermally labile compounds, *Anal. Chem.*, 52, 245, 1980.

96. **Bruins, A.P.**, Probe for the introduction of non-volatile solids into a chemical ionization source by thermal desorption from a platinum wire, *Anal. Chem.*, 52, 605, 1980.

97. **Reinhold, V.N. and Carr, S.A.**, Direct chemical ionization mass spectrometry with poly-imide coated wires, *Anal. Chem.*, 54, 499, 1982.

98. **Reinhold, V.N.**, Direct chemical ionization mass spectrometry of carbohydrates, in *Complex Carbohydrates, Part E, Methods in Enzymology*, Vol. 138, Ginsburg, V., Ed., Academic Press, Orlando, 1987.

99. **McLafferty, F.W.**, Ed., *Tandem Mass Spectrometry*, John Wiley & Sons, New York, 1983.

100. **Busch, K.L., Glish, G.L., and McLuckey, S.A.**, *Mass Spectrometry/Mass Spectrometry. Techniques and Applications of Tandem Mass Spectrometry*, VCH Publishers, New York, 1988.

101. **Cairns, T., Siegmund, E.G., and Stamp. J.J.**, Evolving criteria for the comfirmation of trace level residues in food and drugs by mass spectrometry. Part I, *Mass Spectrom. Rev.*, 8, 93, 1989, Part II, *Mass Spectrom. Rev.*, 8, 127, 1989.

102. **Caldecourt, V.J., Zakett, D., and Tou, J.C.**, An atmospheric pressure ionization mass spectrometer/mass spectrometer, *Int. J. Mass Spectrom. Ion Phys.*, 49, 233, 1983.

103. **Dawson, P.H., French, J.B., Buckley, J.A., Douglas, D.J., and Simmons, D.**, The use of triple quadrupoles for sequential mass spectrometry. I. The instrument parameters, *Org. Mass Spectrom.*, 17, 205, 1982.

104. **Ketkar, S.N., Dulak, J.G., Fite, W.L., Buckner, J.D. and Dheandhanoo, S.**, Atmospheric pressure ionization tandem mass spectrometric system for real-time detection of low-level pollutants in air, *Anal. Chem.*, 61, 260, 1989.

105. **Wysocki, V.H., Kenttämaa, H.I., and Cooks, R.G.**, Internal energy distributions of isolated ions after activation by various methods, *Int. J. Mass Spectrom. Ion Processes*, 75, 181, 1987.

Chapter 4

CHEMICAL IONIZATION REAGENT ION SYSTEMS

I. INTRODUCTION

A wide variety of chemical ionization reagent ion systems have been investigated to date. The following discussion reviews the major systems used in both positive and negative chemical ionization studies and outlines the major types of ionization reactions observed. Detailed considerations of the chemical ionization mass spectra of various classes of compounds are deferred to Chapter 5.

II. POSITIVE ION REAGENT SYSTEMS

A. Brønsted Acid Reagent Systems

By far the most widely used reagent gas systems have been those which yield Brønsted acids (BH^+); these react primarily by proton transfer to species of higher PA than B. Table 1 provides a list of reagent gases and the major reactant ions (BH^+) produced in these gases along with the PAs of the conjugate bases B. The H_3^+ ion, formed by Reaction 1, is the sole product observed in pure hydrogen at high pressures. Reaction 1 is exothermic by ~41 kcal mol^{-1},

$$H_2^{+\cdot} + H_2 \rightarrow H_3^+ + H^\cdot \tag{1}$$

consequently the H_3^+ formed initially is highly excited and is deactivated by collision with the H_2 reagent gas.[1,2] Unless the deactivation process is complete, the exothermicity of proton transfer from H_3^+ will be greater than predicted from the ground state PA data. In 10% X/90% H_2 mixtures (X = N_2, CO_2, N_2O, CO) the XH^+ ion dominates at pressures of 0.3 to 1.0 torr.[3,4] The XH^+ ions are formed by Reactions 1 plus 2 to 4, all of which are known to occur

$$H_3^+ + X \rightarrow XH^+ + H_2 \tag{2}$$

$$H_2^{+\cdot} + X \rightarrow XH^+ + H^\cdot \tag{3}$$

$$X^{+\cdot} + H_2 \rightarrow XH^+ + H^\cdot \tag{4}$$

quite rapidly.[5–7] In the N_2O/H_2 mixture significant ion signals (10 to 15%) are observed for NO^+; this ion is formed by EI of N_2O and does not react with H_2. Methane, the most widely used Brønsted acid reagent gas and that used in the initial study,[8] produces the most complicated spectrum at high pressures. At 1 torr total pressure, the major ions and their approximate relative abundances are CH_5^+ (48%), $C_2H_5^+$ (41%) and $C_3H_5^+$ (6%);[9] there are also minor yields of $C_2H_4^{+\cdot}$ and $C_3H_7^+$. The yields depend to a slight extent on the source pressure and other instrument variables;[10] in the quadrupole ion trap under CI conditions,

71

TABLE 1
Brønsted Acid Chemical Ionization Reagents

Reagent gas	Reactant ion (BH⁺)	PA(B) (kcal mol⁻¹)	HIA(BH⁺) (kcal mol⁻¹)	RE(BH⁺) (eV)
H_2	H_3^+	101.2	300	9.2
N_2/H_2	N_2H^+	118.2	282	8.5
CO_2/H_2	CO_2H^+	130.8	270	7.9
N_2O/H_2	N_2OH^{+a}	138.8	261	7.6
CO/H_2	HCO^+	142.4	258	7.6
CH_4	CH_5^+	131.6	269	7.9
	$C_2H_5^+$	162.6	271	8.3
H_2O	$H^+(H_2O)_n^b$	166.5	234	6.4
CH_3OH	$H^+(CH_3OH)_n^b$	181.9	219	5.7
C_3H_8	$C_3H_7^+$	179.5	250	7.4
$i-C_4H_{10}$	$C_4H_9^+$	195.9	233	6.7
NH_3	$H^+(NH_3)_n^b$	204.0	197	4.8

[a] NO^+ also observed;
[b] Degree of solvation depends on partial pressure of reagent gas. Thermochemical data for monosolvated proton.

the yield of CH_5^+ is considerably reduced and that of $C_2H_5^+$ is correspondingly enhanced.[11] The major primary ions formed in methane upon EI are $CH_4^{+\cdot}$, CH_3^+ and CH_2^+. These react by Reactions 5 to 9 to yield the major ions observed.

$$CH_4^{+\cdot} + CH_4 \rightarrow CH_5^+ + CH_3^{\cdot} \qquad (5)$$

$$CH_3^+ + CH_4 \rightarrow C_2H_5^+ + H_2 \qquad (6)$$

$$CH_2^+ + CH_4 \rightarrow C_2H_4^{+\cdot} + H_2 \qquad (7)$$

$$\rightarrow C_2H_3^+ + H_2 + H^{\cdot} \qquad (8)$$

$$C_2H_3^+ + CH_4 \rightarrow C_3H_5^+ + H_2 \qquad (9)$$

Propane has seen only minor use as a chemical ionization reagent gas. The dominant ion in propane at medium pressures is $C_3H_7^+$ with minor yields of $C_3H_6^{+\cdot}$ and $C_3H_8^{+\cdot}$;[12,13] the primary ions formed by EI of C_3H_8 react with propane primarily by H^- abstraction. At pressures approaching 1 torr, significant yields of $C_5H_{11}^+$ and $C_6H_{13}^+$ are observed.[14] In isobutane at high pressures, $C_4H_9^+$, presumed to be the t-butyl ion, is the dominant ion (>92%) with a minor yield of $C_3H_3^+$ (~3%);[15] again the primary fragment ions react with $i-C_4H_{10}$ by H^- abstraction.[12] The polar reagent gases H_2O, CH_3OH, and NH_3 produce the solvated proton $H^+(B)_n$ with the extent of solvation beyond $n = 1$ depending on the total pressure; typically at 1 torr pressure ions with n in the range 2 to 4 will be observed.[16] For example, Rudewicz and Munson[17] have reported $H^+(NH_3)_3$ (18%), $H^+(NH_3)_2$ (64%), and $H^+(NH_3)$ (18%) in pure ammonia at 0.5 torr and 180°C; by contrast, Hancock and Hodges[18] have reported much lower yields of the more highly solvated species under much the same conditions. To decrease the extent of solvation, it is possible to dilute the polar reagent gas with a nonpolar reagent gas of lower PA such as methane[17] or for ammonia, isobutane. The $CH_3OH^{+\cdot}$, CH_2OH^+, $CH_2O^{+\cdot}$, and CHO^+ primary ions formed in methanol all react relatively rapidly with methanol to form $CH_3OH_2^+$,[19] which subsequently forms more highly solvated species. In both H_2O and NH_3, the following reaction sequence,[20] involving reaction of the primary ions $XH_2^{+\cdot}$ and XH^+ ($X = O$, NH), leads to formation of XH_3^+, which then forms the more highly solvated species.

$$XH_2^{+\cdot} + XH_2 \rightarrow XH_3^+ + XH^{\cdot} \tag{10}$$

$$XH^+ + XH_2 \rightarrow XH_2^{+\cdot} + XH^{\cdot} \tag{11}$$

$$\rightarrow XH_3^+ + X \tag{12}$$

In atmospheric pressure chemical ionization using ambient air carrier gas, the dominant positive reactant ions are solvated protons $H^+(H_2O)_n$ with the range of n depending on the partial pressure of water in the ambient air.[21] The major primary ions produced in air will be $N_2^{+\cdot}$ and $O_2^{+\cdot}$. Good, et al.[22,23] have elucidated the mechanism by which these primary ions are converted to $H^+(H_2O)_n$. For $N_2^{+\cdot}$ the reaction sequence is

$$N_2^{+\cdot} + 2N_2 \rightarrow N_4^{+\cdot} + N_2 \tag{13}$$

$$N_4^{+\cdot} + H_2O \rightarrow H_2O^{+\cdot} + 2N_2 \tag{14}$$

$$H_2O^{+\cdot} + H_2O \rightarrow H_3O^+ + OH^{\cdot} \tag{15}$$

followed by clustering. Neither $O_2^{+\cdot}$ nor $O_4^{+\cdot}$ are capable of undergoing charge transfer to H_2O and the conversion proceeds mainly as follows:

$$O_2^{+\cdot} + 2O_2 \rightarrow O_4^{+\cdot} + O_2 \tag{16}$$

$$O_2^{+\cdot} + H_2O + O_2 \rightarrow O_2^{+\cdot}(H_2O) + O_2 \tag{17}$$

$$O_4^{+\cdot} + H_2O \rightarrow O_2^{+\cdot}(H_2O) + O_2 \tag{18}$$

$$O_2^{+\cdot}(H_2O) + H_2O \rightarrow H_3O^+OH + O_2 \tag{19}$$

$$H_3O^+OH + H_2O \rightarrow H^+(H_2O)_2 + OH^{\cdot} \tag{20}$$

The major reaction mode of the Brønsted acids, BH^+, is by proton transfer (Reaction 21) to give initially the protonated molecule MH^+. As the data in

$$BH^+ + M \rightarrow MH^+ + B \qquad (21)$$

Chapter 2, Section III.B show, this reaction usually occurs with high efficiency provided it is exothermic, i.e., provided PA(M) > PA(B). Examination of the PA data (Tables 17 to 20, Chapter 2) indicates that $C_3H_7^+$ and reagent ions higher in Table 1 than $C_3H_7^+$ will react by exothermic proton transfer to most organic molecules. By contrast, $C_4H_9^+$ and $H^+(NH_3)_n$ are more specific protonating agents, particularly the latter ion. In general $H^+(NH_3)_n$ will protonate organic compounds with PA > ~205 kcal mol^{-1}; in practice this means that primarily nitrogen-containing compounds will be protonated. With compounds having PA < ~205 kcal mol^{-1} the ammonia CI mass spectra frequently show [M + NH_4]$^+$ adduct ions. In a systematic study, Keough and DeStefano[24] showed that [M + NH_4]$^+$ was not formed to a significant extent if the PA of M was less than ~188 kcal mol^{-1} and that a lone pair of electrons at the basic site is a necessary prerequisite for formation of the adduct ion. Rudewicz and Munson[17] showed that the yield of [M + NH_4]$^+$ ions from oxygenated compounds increased as the ammonia reagent was diluted with methane and attributed this to a decrease in the rate of the ligand-switching reaction

$$\left[M + NH_4\right]^+ + NH_3 \rightarrow NH_4^+NH_3 + M \qquad (22)$$

as the ammonia was diluted. Reaction 22 has been observed directly by Cody [25] using a two-cell FTMS.

In the ammonia chemical ionization of a number of substituted benzenes, C_6H_5X, a substitution reaction occurs to give $C_6H_5NH_3^+$.[26] Under the low pressure conditions prevailing in an ion cyclotron resonance spectrometer, it was shown[27] for chlorobenzene that the reaction forming this product was not

$$NH_4^+ + C_6H_5Cl \rightarrow C_6H_5NH_3^+ + HCl \qquad (23)$$

but rather was reported to be both of

$$C_6H_5Cl + NH_3^{+\cdot} \rightarrow C_6H_5NH_3^+ + Cl^{\cdot} \qquad (24)$$

$$C_6H_5Cl^{+\cdot} + NH_3 \rightarrow C_6H_5NH_3^+ + Cl^{\cdot} \qquad (25)$$

More recent work[28] has shown that Reaction 24 does not occur, but that $NH_3^{+\cdot}$ undergoes charge transfer with C_6H_5Cl which then reacts by Reaction 25. By contrast, under high pressure conditions similar to those prevailing in CI, it has been reported[29] that Reaction 23 does occur, so that the exact origin of the substitution product remains in doubt. A variety of alcohols also undergo a substitution reaction in ammonia CI,[26] the reaction being nominally

$$NH_4^+ + ROH \rightarrow RNH_3^+ + H_2O \qquad (26)$$

Depending on the system S_N2-like, S_N1-like, and S_Ni mechanisms have been implicated, as discussed in detail by Westmore and Alauddin.[26] The overall reaction gives a product with the same m/z ratio as the molecular ion of the hydroxy compound.

Ammonia and other polar reagent gases will tend to solvate protonated molecules to form $MH^+(B)$. When such clustering reactions are undesirable Barry and Munson[30] have proposed the use of diisopropyl ether in N_2, CH_4, or He as a reagent. The ether has a PA similar to that of ammonia, and the dominant ion, the protonated ether, does not readily form cluster ions. Alkyl amines give Brønsted acid reagent ions which are even milder protonating agents than NH_4^+.[31] Other cluster ions frequently observed (usually in low abundance) are $[M + C_2H_5]^+$ and $[M + C_3H_5]^+$ in CH_4 CI mass spectra, $[M + C_3H_7]^+$ in propane CI mass spectra and $[M + C_3H_3]^+$ and $[M + C_4H_9]^+$ in i-butane CI mass spectra.

In principle, the BH^+ ions also can react by hydride abstraction (Reaction 27) or by charge exchange (Reaction 28).

$$BH^+ + M \rightarrow [M - H]^+ + BH_2 \left(or\ B + H_2\right) \qquad (27)$$

$$BH^+ + M \rightarrow M^{+\cdot} + BH \left(or\ B + H^\cdot\right) \qquad (28)$$

Accordingly, we have given in Table 1 the effective HIAs and REs of the BH^+ reactant ions. The HIAs should be considered in light of the values given in Table 8, Chapter 2. The H_3^+ ion has a high HIA and in most H_2 CI mass spectra, appreciable $[M - H]^+$ ion signals are observed. In addition, ground state H_3^+ has a RE of ~9.3 eV and is capable of undergoing charge exchange with any molecule having an IE lower than this value; consequently, $M^{+\cdot}$ ions are seen in many H_2 CI mass spectra. The N_2H^+ ion is a moderately strong hydride ion abstracting reagent and is capable of undergoing charge exchange with species having IEs below ~8.6 eV. The CO_2H^+, N_2OH^+ and HCO^+ reagent ions are weaker hydride ion abstracting reagents but still are capable of abstracting secondary and tertiary alkyl hydrogens, allylic and benzylic hydrogens, and those alpha to hydroxyl groups. These reactant ions are unlikely to react to any significant extent by charge exchange. It should be noted that 10 to 15% of the ion yield in N_2O/H_2 mixtures resides in NO^+, and the characteristic reactions of this ion (Section II.E) are also observed when using this reagent gas system.

The CH_5^+ and $C_2H_5^+$ ions produced in methane are capable of abstracting hydride from many classes of molecules, and $[M - H]^+$ ions frequently are observed in methane CI mass spectra. The $C_2H_5^+$ ion has a RE of ~8.5 eV and will charge exchange only with molecules of quite low IE. However, the methane reagent ion system includes a low abundance $C_2H_4^+$ ion (RE = 9 to 11 eV)[32] which can charge exchange with many molecules to produce low abundance (usually ≤1%) molecular ions of the additive. The $C_3H_7^+$ ion is capable of

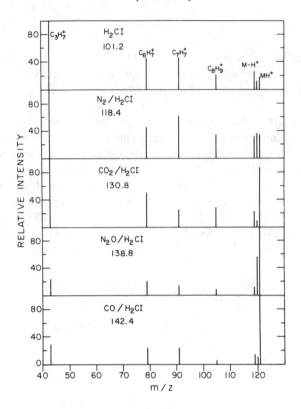

FIGURE 1.   Effect of reagent PA on Brønsted acid CI mass spectra of n-propylbenzene. Numbers give PA of B in BH⁺ reagent. Data from Herman and Harrison.[33]

abstracting hydride ion from a variety of molecules but the remaining reagent ions are very weak hydride ion abstracting reagents; they also have very low REs and are unlikely to react by charge exchange.

For the proton transfer Reaction 21, $\Delta H^0 = PA(B) - PA(M)$. The exothermicity of the proton transfer reaction resides largely in the $M - H^+$ bond formed[3] and is available as internal energy to promote fragmentation of the MH^+ ion. To maximize the probability of MH^+ ion formation, which provides molecular mass information, one utilizes the least exothermic protonation reaction possible. On the other hand, structural information is derived from the fragment ions observed in the CI mass spectrum. The extent of fragmentation depends on the identity of M and on the protonation exothermicity. Both effects are illustrated by the proton transfer CI mass spectra shown in Figures 1 and 2. For both figures, the protonation exothermicity decreases as one moves down the figure and it is clear that the extent of fragmentation decreases and the MH^+ ion intensity increases as the protonation exothermicity decreases. Equally clearly, the oxime (Figure 2) shows a much greater tendency for the MH^+ ion to fragment. It frequently is observed that CH_4 CI mass spectra combine an adequate MH^+ ion signal for molecular mass determination with

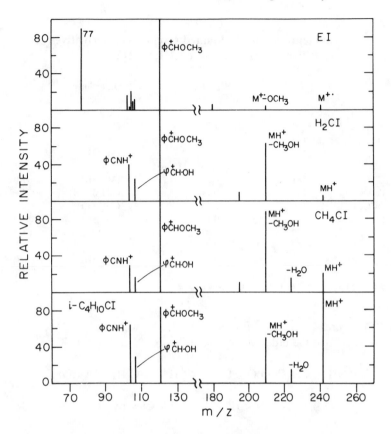

FIGURE 2.   Effect of protonation exothermicity on Brønsted acid CI mass spectra of benzoin methyl ether oxime. Data from Harrison and Kallury.[34]

appropriate fragmentation for structure elucidation; consequently, CH_4 is the most useful single Brønsted acid reagent gas.

The fragmentation reactions of the even-electron MH^+ ion usually involve elimination of an even-electron neutral (a stable molecule) to form an even-electron fragment ion. If the molecule contains a functional group Y, the initial fragmentation frequently involves loss of the stable molecule HY from MH^+; note, as an example, the loss of H_2O and CH_3OH from MH^+ in the spectra of Figure 2. The tendency of the functional group Y to be eliminated as HY is inversely related to the PA of HY.[35] This can be rationalized by noting that the reverse of Reaction 29 is the addition of R^+ to HY. If R^+ was a proton the

$$RYH^+ \rightarrow R^+ + HY \tag{29}$$

exothermicity of the reverse reaction would be the PA of HY. Thus, it is reasonable to project that the endothermicity (and, hence, the critical reaction energy) of the forward reaction should be proportional to PA(HY) and that, as

TABLE 2
Fragmentation of Protonated Cyclohexyl Derivatives
$$c\text{-}C_6H_{11}YH^+ \rightarrow c\text{-}C_6H_{11}^+ + HY$$

Y	PA(HY)	$MH^+/C_6H_{11}^+$	Other ions[a]	
SCOCH$_3$	—	1.7	CH$_3$COSH$_2^+$	(73%)
NH$_2$	204	1.7	[M – H]$^+$	(70%)
NCO	173	1.3	—	
PH$_2$	189	1.0	[M – H]$^+$	(38%)
NCS	—	0.65	—	
C$_6$H$_5$	181	0.23	[M – H]$^+$	(23%)
OCOC$_2$H$_5$	192	0.17	C$_2$H$_5$CO$_2$H$_2^+$	(310%)
OCOCH$_3$	188	0.15	CH$_3$CO$_2$H$_2^+$	(84%)
SH	170	0.14	—	
OCOH	179	0.07	HCO$_2$H$_2^+$	(13%)
OH	167	0.03	—	
F, Cl, Br, I	~140	0.00	[M – H]$^+$	(1–7%)

[a] Expressed as % MH$^+$ or C$_6$H$_{11}^+$, whichever is larger.

PA(HY) decreases, a greater fraction of the RYH$^+$ will fragment. Accordingly, one predicts that the order of the ease of loss of Y as HY should be

$$NH_2 < CH_3C(:O)O < CH_3S < CH_3O < C_6H_5 \approx HC(:O)O$$

$$< CN < HS < OH < I \approx Cl \approx Br.$$

There is some experimental support for this inverse dependence of the extent of fragmentation of RYH$^+$ on PA(HY).[36,37] Table 2 shows the MH$^+$/C$_6$H$_{11}^+$ intensity ratios observed[36] in the CH$_4$ CI mass spectra of cyclohexyl derivatives (C$_6$H$_{11}$Y). There is a general correlation with PA(HY), although, when other fragmentation routes are important, as, for example, the formation of protonated acid ions from cyclohexyl esters, the correlation begins to break down.

While this approach suffices to rationalize the relative extent of fragmentation of related monofunctional compounds, the task of predicting the fragmentation of more complex, polyfunctional, compounds is much more difficult. One approach which has been used is to assume that protonation at a given site or functionality triggers fragmentation at that site. As an example, it was assumed[38] in early studies of the CH$_4$ CI of amino acids that protonation at the hydroxyl group resulted in elimination of water, protonation at the carbonyl oxygen led to elimination of formic acid, while protonation at the amine group led to elimination of ammonia (Scheme 1). However, isotopic labelling studies[39] have shown that the added proton interchanges between the various basic sites prior to fragmentation of MH$^+$, casting doubt on the localized activation model in this case. However, there are examples of specific site

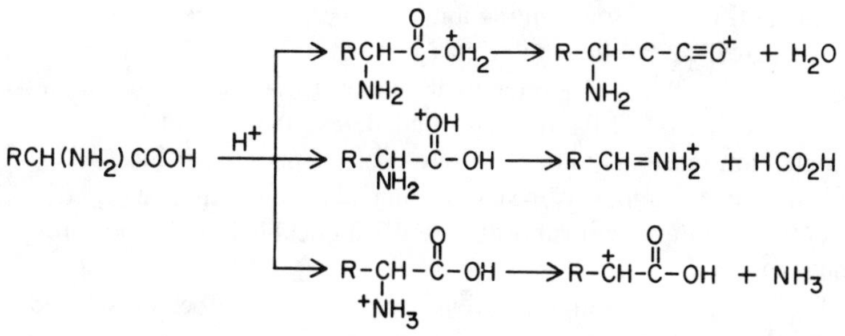

SCHEME 1

protonation triggering fragmentation. Mason et al.[40] have shown that fluorobenzene is protonated, in part, at the fluorine and that only this species loses HF on collisional activation. More recently, it has been shown[41] that protonation of bifunctional compounds of the type p-XCH$_2$C$_6$H$_4$CH$_2$Y (X = NH$_2$, NH(CH$_3$), or N(CH$_3$)$_2$; Y = OH or OCH$_3$) occurs primarily at X and Y with specific loss of HX and HY from the species protonated at X and Y, respectively.

An alternative approach, suggested by detailed studies of several classes of compounds,[3,4,33,42,43] is to assume a competitive kinetic scheme for fragmentation of MH$^+$ similar to the quasi-equilibrium theory used to rationalize EI mass spectra.[44,45] In this approach, the initial protonation reaction is assumed to produce a collection of MH$^+$ ions with an average excess internal energy determined by the exothermicity of the protonation reaction. These excited ions, MH^{+*}, are assumed to be sufficiently long-lived that the excess energy is randomized among the internal degrees of freedom. The MH^{+*} ions may undergo collisional de-excitation (Reaction 30) or alternatively, may fragment by one of several competing fragmentation reactions (Reactions 31 and 32).

$$MH^{+*} + B \rightarrow MH^+ + B \qquad (30)$$

$$MH^{+*} \rightarrow F_1^+ + N_1 \qquad (31)$$

$$MH^{+*} \rightarrow F_2^+ + N_2 \qquad (32)$$

For each fragmentation reaction the rate coefficient, k(E), can be calculated by suitable application of the absolute rate theory and, in its simplest form, can be approximated by[44,45]

$$k(E) = \upsilon \left\{ \left(E - \varepsilon_0 \right) / E \right\}^s \qquad (33)$$

where υ is an energy-independent term or frequency factor, E is the excess internal energy of the fragmenting ion, ε_0 is the critical energy for fragmentation (often called the activation energy) and s is the effective number of oscillators, often taken as $\sim^1/_3$ the total number (3N-6). At the usual pressures of the bath gas (0.5 to 1.0 torr), collisional de-excitation is relatively rapid and fragmentation reactions must have rate coefficients greater than $\sim 10^7$ s^{-1} for fragmentation to compete with stabilization. (This minimum rate coefficient is greater than the minimum value of $\sim 10^5$ s^{-1} applicable in EI-induced fragmentation.)

The factors determining reaction competition have been discussed elsewhere[44,45] for EI-induced fragmentations; the basic conclusions are applicable here. In general, relative rate coefficients for competing fragmentation reactions are determined primarily by the relative critical energies at low excess internal energies and by the relative frequency factors at high internal energies. The frequency factor is lower for a fragmentation process involving rearrangement (a tight transition state) than it is for a simple bond fission reaction (loose transition state). It appears that the relative critical energies, which frequently can be related to reaction enthalpies, play a major role in CI-induced fragmentation reactions. Thus, in many cases, the relative importance of fragmentation reactions can be deduced if the relative endothermicities for fragmentation of ground state MH$^+$ can be estimated. Table 3 illustrates this for the two major competing fragmentation Reactions 34 and 35, for protonated butyl acetates.[42]

$$CH_3CO_2C_4H_9 \cdot H^+ \rightarrow CH_3C(OH)_2^+ + C_4H_8 \tag{34}$$

$$\rightarrow C_4H_9^+ + CH_3CO_2H \tag{35}$$

For t-butyl acetate formation of t-C$_4$H$_9^+$ is thermochemically favored and C$_4$H$_9^+$ is the dominant fragment ion. In contrast, for s-butyl acetate, formation of CH$_3$C(OH)$_2^+$ is thermochemically favored and this species forms the base peak in the spectrum. It is quite likely that the reaction actually proceeds through the intermediacy of an ion/neutral complex[46] consisting of a C$_4$H$_9^+$ ion complexed with an acetic acid molecule, which may change some of the critical reaction energies to some degree but the basic arguments remain unchanged. In this ion/neutral complex, rearrangement of the alkyl ion to a more stable structure can readily occur and the results for i-butyl acetate in Table 3 can be rationalized only by assuming that the butyl ion has rearranged to the more stable t-butyl structure. Reaction 34 is a rearrangment reaction and should have a lower frequency factor (tighter transition state) than the bond fission Reaction 35. In such a case, one would expect Reaction 35 to be favored at higher internal energies and the C$_4$H$_9^+$/CH$_3$C(OH)$_2^+$ ratio in s-butyl acetate is observed to be 0.13 in the CO/H$_2$ CI mass spectrum but increases to 0.62 in the more exothermic H$_2$ CI system.

Even when the reaction thermochemistry cannot be calculated explicitly, the thermochemically favored reaction frequently can be deduced from a knowledge of the factors which determine relative ion and neutral stabilities.

TABLE 3
Relative Fragment Ion Intensities in CO_2/H_2 Chemical Ionization of Butyl Acetates (CH_3CO_2R)

R	Relative intensity $CH_3C(OH)_2^+$	ΔH^{0a} (kcal mol^{-1})	Relative intensity $C_4H_9^+$	ΔH^{0a} (kcal mol^{-1})
n-C_4H_9	100	24	31	52 (32)[b]
i-C_4H_9	89	22	100	50 (17)[c]
s-C_4H_9	100	26	22	36
t-C_4H_9	9	29	100	23

[a] ΔH^0 for reaction of ground state $CH_3CO_2R \cdot H^+$.
[b] Value in parentheses for s-$C_4H_9^+$ structure.
[c] Value in parentheses for t-$C_4H_9^+$ structure.

Two examples will suffice as illustrations. The ion A formed by elimination of the elements of formic acid from protonated amino acids is greatly stabilized by the amino group and constitutes the most abundant fragment ion in the CH_4 CI mass spectra of many amino acids.[38,39] By contrast, the [MH$^+$–NH$_3$] ion B is destabilized by the adjacent carbonyl group and loss of NH$_3$ is observed only when the critical reaction energy is lowered by anchimeric assistance by a neighboring group as in formation of C from protonated methionine.

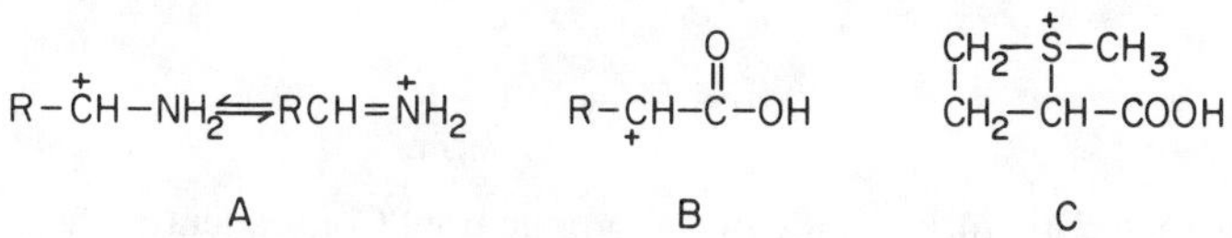

A   B   C

SCHEME 2

As a second example we note that considerations of the relative PAs of H_2O and CH_3OH would suggest that H_2O loss should be preferred over CH_3OH loss from the protonated oxime of Figure 2. However, the benzylic ion formed by CH_3OH loss is more stable than the ion formed by H_2O loss and [MH$^+$–CH$_3$OH] is the more abundant ion in the CI mass spectra.

Finally, we should note that, while formation of an even-electron ion by elimination of an even-electron neutral from MH$^+$ normally is the thermochemically favored fragmentation route, there are a few examples where formation of an odd-electron fragment ion by elimination of an odd-electron neutral (radical) is the favored mode of fragmentation. For example, the CH_4 CI mass spectra of iodobenzene and the iodotoluenes show[47] abundant ion signals corresponding to [MH$^+$–I$^·$] and no signal corresponding to [MH$^+$–HI]. Similarly, the CH_4 CI mass spectra of the bromoanisoles and bromoanilines show[48] extensive fragmentation by loss of a bromine atom from MH$^+$ but little loss of HBr. In these cases, the high heat of formation of the appropriate phenyl cation relative to the appropriate odd-electron benzene molecular ion makes Reaction 36 energetically more demanding than Reaction 37.

$$RC_6H_4X^-H^+ \rightarrow RC_6H_4^+ + HX \tag{36}$$

$$\rightarrow RC_6H_5^{+\cdot} + X^{\cdot} \tag{37}$$

B. Hydride Ion Abstraction Reagent Systems

There have been no specific reagent systems developed which involve hydride ion abstraction as the major or sole ionization reaction. The NO^+ ion reacts with some classes of compounds primarily by H^- abstraction but also shows other reaction modes; the use of NO as a reagent gas is discussed in Section II.E. The $C_2H_5^+$ ion produced in methane reacts with alkanes entirely by H^- abstraction[49] but usually reacts predominantly by proton transfer when the PA of the substrate molecule makes proton transfer exothermic. The data in Table 8, Chapter 2 show that CF_3^+ has a high HIA, while the data summarized in Table 6, Chapter 2 show that CF_3^+ reacts moderately rapidly with alkanes by H^- abstraction; these results suggest that CF_3^+ might make an efficient H^- abstraction reagent ion. The one brief report[50] does not bear out this prediction.

C. Charge Exchange Reagent Systems

An alternative mode of ionization which has seen some use is that of charge exchange, Reaction 38. In this mode, the odd-electron $M^{+\cdot}$ molecular

$$R^{+\cdot} + M \rightarrow M^{+\cdot} + R \tag{38}$$

ion, characteristic of EI, is the initial product ion. Consequently, the fragmentation reactions which are observed are the same as those observed in EI mass spectrometry. The essential difference is that in EI, the $M^{+\cdot}$ ions have a distribution of internal energies[44,45] whereas in charge exchange, the $M^{+\cdot}$ ions have a discrete internal energy or a small range of internal energies determined by the exothermicity of Reaction 38. Consequently, the EI mass spectrum is the net result of summing an energy-dependent competitive kinetic scheme over a distribution of internal energies while the charge exchange mass spectrum is the net result of the energy-dependent scheme operating at a fixed internal energy or over a small range of internal energies. The exothermicity of Reaction 38 is determined by the RE of the reactant ion less the ionization energy (IE) of the additive M; the small range of internal energies will arise if not all of the exothermicity ends up as internal energy or if the reactant ions have more than one RE. Table 4 provides a partial list of possible charge exchange reagent systems with the REs of the dominant ion. Since most organic molecules have IEs in the range of 8 to 12 eV, a wide range of internal energies can be imparted to the molecular ions by varying the reactant ion. The effect of varying the reaction exothermicity on the charge exchange mass spectrum is shown by the data in Table 5 for cyclohexane (IE = 9.88 eV).[51] The results clearly show the distinct difference between EI and charge exchange mass spectra and the

TABLE 4
Charge Exchange Reagent Systems

Reagent gas	Reactant ion	RE (eV)
C_6H_6	$C_6H_6^{+\cdot}$	9.3
N_2/CS_2	CS_2^+	~10.0
CO/COS	$COS^{+\cdot}$	11.2
Xe	$Xe^{+\cdot}$	12.1, 13.4
CO_2	CO_2^+	13.8
CO	$CO^{+\cdot}$	14.0
Kr	$Kr^{+\cdot}$	14.0, 14.7
N_2	$N_2^{+\cdot}$	15.3
Ar	$Ar^{+\cdot}$	15.8, 15.9

strong dependence of the charge exchange mass spectrum on the RE of the reactant ion.

The rare gas ions, $N_2^{+\cdot}$, CO_2^+, and CO^+ can be produced by EI of the appropriate pure gas at high source pressures; at such pressures, some formation of dimer ions will occur. These dimer ions will have REs which are lower than the monomer ions by the binding energy in the dimeric species. For example, this binding energy is about 1 eV for the $N_2^{+\cdot}$–N2 dimer.[52] The chlorobenzene molecular ion is the dominant ion (>85%) in pure chlorobenzene at high pressure[53] while the benzene molecular ion has been prepared by ionization of a 20% C_6H_6/80% He mixture[54] and the $C_6F_6^+$ ion is the dominant species (>95%) in a 10% C_6F_6/90% CO_2 mixture at high pressures.[55] The N_2O^+ reagent ion has been prepared[55] by ionization of a 10% N_2O/90% N_2 mixture. The CS_2^+ ion has been produced[56-58] by EI of N_2/CS_2 mixtures where, at high pressures, CS_2^+ is the major species (>80%); minor yields of $S_2^{+\cdot}$, $(CS_2)_2^+$, and CS_3^+ are also observed.[58] $COS^{+\cdot}$ has been produced[58] in a similar manner by ionization of a COS/CO mixture at high pressures; minor yields of $S_2^{+\cdot}$, CS_2^+, and $(COS)_2^{+\cdot}$ also were observed.

Charge exchange chemical ionization has been used to explore the energy dependence of the fragmentation of odd-electron molecular ions;[51,54,58,59] typical results are presented in Table 5 and show the energy evolution of the ions seen in the EI mass spectrum. From an analytical viewpoint, a number of studies[54,60,61] have shown that low energy charge exchange enhances the differences in the mass spectra of isomeric molecules compared to the differences observed in the 70 eV EI mass spectra; the use of charge exchange has the advantage, compared to low energy EI,[62] that the sensitivity of ionization is maintained. This use of charge exchange is illustrated in Figure 3 which compares the 70 eV EI mass spectra and the $C_6F_6^+$ charge exchange mass spectra of *cis*-and *trans*-4-methylcyclohexanol. Not only are the charge exchange mass spectra much simpler but also the mass spectral differences for the two isomers are much enhanced.

A second application of charge exchange chemical ionization is for the

TABLE 5
Electron Ionization and Charge Exchange Mass Spectra of Cyclohexane

Ionization	Fractional intensities						
	$C_6H_{12}^{+\cdot}$	$C_5H_9^+$	$C_4H_8^{+\cdot}$	$C_4H_7^+$	$C_3H_7^+$	$C_3H_6^{+\cdot}$	$C_3H_5^+$
EI (70 eV)	0.26	0.07	0.31	0.10	0.04	0.08	0.15
CE(CS_2^+)	1.0	—	—	—	—	—	—
CE(COS^+)	0.91	0.03	0.06	—	—	—	—
CE(Xe^+)	0.03	0.15	0.60	0.11	—	0.10	0.01
CE(CO^+)	0.08	0.05	0.41	0.22	—	0.08	0.15
CE(N_2^+)	0.06	0.01	0.12	0.27	—	0.07	0.49

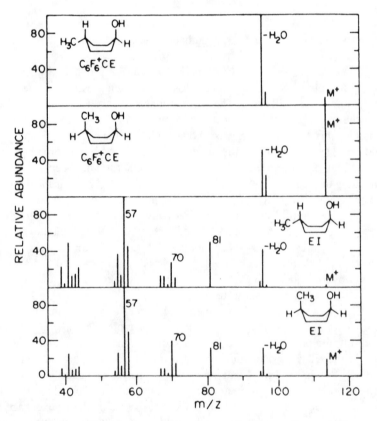

Figure 3.   Electron ionization and $C_6F_6^+$ charge exchange mass spectra of *cis*- and *trans*-4-methylcyclohexanol. (From Harrison, A.G. and Lin, M.S., *Org. Mass Spectrom.*, 19, 67, 1984. With permission.)

selective ionization of specific components of mixtures. The principle of the method is illustrated in Figure 4 taken from the work of Sieck.[53] Figure 4 plots the IEs of alkanes, cycloalkanes, alkylbenzenes, and alkylnaphthalenes (components of petroleum products) as a function of molecular weight. Also in-

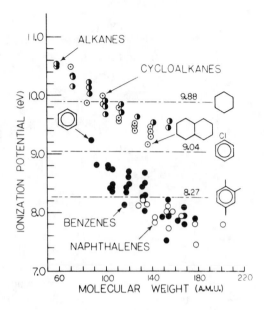

Figure 4. Ionization potentials of hydrocarbons as a function of molecular weight and type. (From Sieck, L.W., *Anal. Chem.* 55, 38, 1983. With permission)

cluded as horizontal lines on the diagram are the REs (taken to be numerically the same as the IE of the neutral) of c-$C_6H_{12}^{+ \cdot}$, $C_6H_5Cl^{+ \cdot}$, and $(CH_3)_3C_6H_3^{+ \cdot}$. These species should ionize only those hydrocarbons with IEs below the appropriate line. Clearly, charge exchange ionization of a mixture using $C_6H_5Cl^{+ \cdot}$ as the reagent ion will ionize only the benzene and naphthalene components while ionization using $(CH_3)_3C_6H_3^{+ \cdot}$ will be even more selective for naphthalenes and the higher molecular weight benzenes.

Selective ionization by charge exchange was first demonstrated by Subba Rao and Fenselau[63] who used $C_6H_6^{+ \cdot}$ (benzene) for selective ionization of esters of unsaturated fatty acids in the presence of saturated fatty acid esters. A similar use of $C_6H_6^{+ \cdot}$ for selective ionization of aromatic hydrocarbons in hydrocarbon mixtures has been suggested by Sunner and Szabo,[64] while Munson and co-workers[55] also have demonstrated the use of $C_6H_6^{+ \cdot}$ for selective ionization. Charge exchange from $C_6H_6^{+ \cdot}$ has been used for selective ionization of polychlorinated biphenyls (PCBs) in atmospheric pressure chemical ionization experiments.[65]

D. Condensation Reaction Reagents

Condensation ion/molecule reactions have not found extensive applications in chemical ionization mass spectrometry. A method of locating double bonds without derivatization which involves a condensation ion/molecule reaction between vinyl methyl ether (VME) and the olefinic compound has been developed by Ferrer-Correia et al.[66,67] The principle involved is illustrated

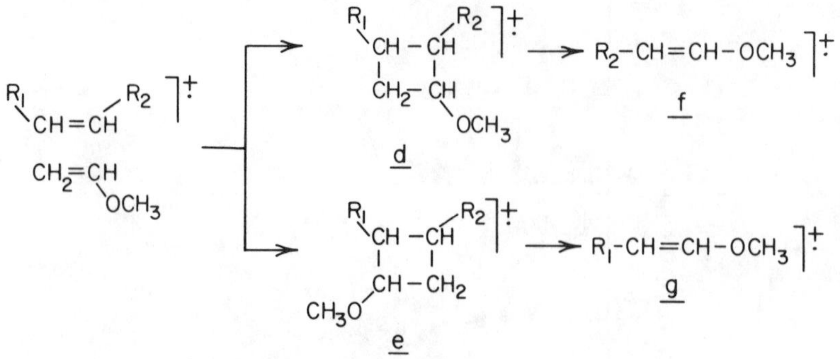

SCHEME 3

in Scheme 3. Four-centered addition complexes D and E are formed by attack of either olefinic molecular ion on the second neutral olefin; these fragment, in part, to characteristic ionic products F and G, the m/z ratios of which identify the size of R_1 and R_2 and, hence, the position of the double bond. Alternative fragmentation reactions of the complexes involve loss of CH_3 and loss of CH_3OH. It is not known which olefinic moiety is ionized and, quite likely, both molecular ions are capable of reacting with the other neutral.

In the early studies[66,67] either pure VME or VME/CO_2 mixtures were used as the reagent gas. Ionization of pure VME produces a complex array of ions as does ionization of the 10% VME/90% CO_2 mixtures where charge exchange from the primary $CO_2^{+\cdot}$ ions to either VME or the added olefin results in extensive fragmentation. Considerably simpler spectra are obtained if a reagent gas mixture containing approximately 75% N_2, 20% CS_2, and 5% VME is used.[68,69] In this latter mixture, the initial EI produces mainly $N_2^{+\cdot}$ which, by charge exchange, forms $CS_2^{+\cdot}$. The $CS_2^{+\cdot}$ ion is a low energy charge exchange reagent ion (Table 4), and on reaction with either VME or the olefinic compound produces predominantly molecular ions. The method has achieved only very limited success in practical applications;[70] for example, it has been shown to be unsatisfactory for double bond location in polyunsaturated acids.[71]

E. NO as a Positive Ion Reagent System

EI of either pure NO or an N_2/NO mixture in a conventional CI source produces NO^+ in good yield with lesser amounts of the dimeric ion $(NO)_2^{+\cdot}$. The use of this reagent ion was first explored by Hunt and Ryan.[72] In general, three alternative reaction channels, addition (Reaction 39), hydride ion abstraction (Reaction 40) and charge exchange (Reaction 41) are observed depending on

$$NO^+ + M \rightarrow M \cdot NO^+ \tag{39}$$

$$\rightarrow [M - H]^+ + HNO \tag{40}$$

$$\rightarrow M^{+\cdot} + NO \cdot \tag{41}$$

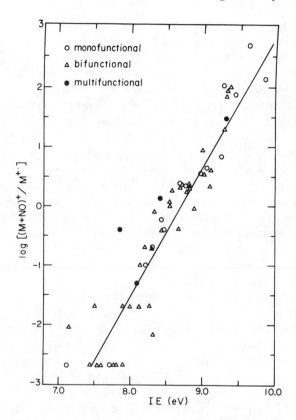

FIGURE 5. Plot of log ($[M + NO]^+/M^{+\cdot}$) vs. IE of substituted benzene in NO^+ CI of benzenes. Data from Daishima et al.[74]

the substrate molecule. In addition, in some cases, hydroxide ion abstraction from alcohols or halide ion abstraction from alkyl halides is observed. The IE of NO is 9.3 eV,[73] but the effective RE of NO^+ under CI conditions appears to be considerably lower. With many aromatic molecules, NO^+ reacts either by charge exchange (Reaction 41) or by adduct ion formation (Reaction 39). Figure 5 shows a plot of log ($[M + NO]^+/M^{+\cdot}$) vs. the IE of the aromatic using the data recently reported by Daishima et al.;[74] results where significant other reaction channels were observed have been omitted from the plot. An approximately linear relation is observed. From this plot, one can conclude that equal yields of $[M + NO]^+$ and $M^{+\cdot}$ will be observed for aromatic molecules with IEs of ~8.7 eV. For aromatic molecules with higher IEs adduct ions will be favored, while for aromatic molecules with lower IEs formation of $M^{+\cdot}$ will be favoured. This conclusion, which seems to apply roughly to other classes of molecules, is in agreement with earlier more qualitative results.[75,76] The results suggest that, under CI conditions, NO^+ ions have a narrow range of effective REs ~8.7 eV.

The HIA of NO^+ is ~246 kcal mol^{-1} with the result that NO^+ is capable of abstracting, as hydride ion, secondary or tertiary alkyl hydrogens, secondary allylic hydrogens, benzylic hydrogens, and hydrogens bonded to the same carbon as hydroxyl groups (see Table 8, Chapter 2). In general, NO^+ reacts with aldehydes and alkyl ethers to give $[M - H]^+$, with ketones to give $[M + NO]^+$ and with alkyl esters to give $[M + NO]^+$ with a minor yield of RCO^+ ions.[75,76] With primary and secondary alcohols, $[M - H]^+$ ions dominate the spectra while for tertiary alcohols, $[M - OH]^+$ ions are the major products.[76,77] With internal olefins, NO^+ reacts by hydride abstraction and by NO^+ addition or charge exchange depending on the IE of the olefin. With terminal olefins, an additional reaction mode involves addition of NO^+ to the double bond followed by elimination of small neutral alkane molecules to give a series of ions $C_nH_{2n}NO^+$ ($n = 3$ to 6),[70,77,78] with the intensity shifting to larger values of n with increasing chain length. Long-chain olefins with internal double bonds, as well as a number of olefinic derivatives and alkynes, also undergo NO^+ addition to the multiple bond and yield nitrogen-containing ions which provide information on the position of the double or triple bond;[70,77] the intensity of these products, however, is very dependent on experimental conditions, particularly the source temperature.

There are several problems associated with the use of NO as a reagent gas. It has been reported[78] that the lifetime of a rhenium filament in the presence of 1 torr of NO is about 6 h. In addition, depending on experimental conditions, there may be oxidation of the substrate by reactions which are not clearly understood. Thus alkanes, in addition to forming $[M - H]^+$ ions, may show $[M - 3H]^+$ and $[M - 2H + NO]^+$ ions resulting from oxidation of the alkane to an alkene and ionization of the alkene by NO^+. Primary alcohols may show $[M - 3H]^+$ and $[M - 2H + NO]^+$ ions apparently resulting from ionization of the aldehyde formed by oxidation of the alcohol. In the same way secondary alcohols may be oxidized to ketones and show $[M - 2H + NO]^+$ from ionization of this ketone. Recently, it has been reported[79,80] that olefinic compounds of the type $CH_3(CH_2)_xCH=CH(CH_2)_yR$ ($R = CH_2OH$, CH_2OAc, or CHO) give abundant acyl ions at low sample pressures and source temperatures which identify the position of the double bond. Although the detailed mechanism of formation of these ions is not clear,[81] there is some evidence that a neutral surface-catalyzed oxidative cleavage of the double bond results in the formation of the appropriate aldehydes ($CH_3(CH_2)_xCHO$ and $R(CH_2)_yCHO$) which are ionized by hydride abstraction by NO^+. Although these various oxidation reactions potentially provide more analytical information, the dependence on experimental parameters limits their usefulness. Chai and Harrison[76] have reported that a 5% $NO/95\%$ H_2 mixture at 0.2 to 0.3 torr yields NO^+ as the dominant reactant ion and that the oxidation reactions of alcohols and alkanes do not occur under these conditions. The major reactions forming NO^+ in the NO/H_2 mixture are[82]

$$H_3^+ + NO \rightarrow HNO^+ + H_2 \tag{42}$$

$$HNO^+ + NO \rightarrow NO^+ + HNO \tag{43}$$

In the same vein, it has been reported[83] that the use of nitromethane as a reagent gas gives a substantial yield of NO^+ among the reactant ions and CI mass spectra similar to those obtained with NO as the reagent gas. Polley and Munson[84] have reported the use of N_2O as a reagent gas; both NO^+ and $N_2O^{+\cdot}$ are observed at CI pressures, the latter acting as a charge exchange reagent ion (see Section C).

F. Tetramethylsilane (TMS) Reagent

Vouros and co-workers[85–87] observed that tetramethylsilane (TMS) at CI pressures gave predominantly $(CH_3)_3 Si^+$ as a reagent ion and that this ion reacted with compounds containing n- or π-electrons (Lewis bases) to give [M + $(CH_3)_3Si]^+$ ([M + 73]$^+$) adduct ions. In similar experiments, Stillwell et al.[88] showed that 5:1 mixtures of TMS with isobutane or nitrogen gave mainly the trimethylsilyl ion which gave adduct ions with a variety of nonvolatile compounds using a direct exposure probe to introduce the sample; the adduct ions were identified by using equal amounts of $(CH_3)_4Si$ and $(CD_3)_4Si$ in the reagent mixture which resulted in two adduct ions separated by 9 amu.

More recently, Clemens and Munson[89] have shown that a 1:1 mixture of TMS and methane, to which a small amount of a base B (such as H_2O, $(CH_3)_2CO$, or $(CH_3)_2O$) was added, yielded $(CH_3)_3Si\,B^+$ as the dominant ion. This species reacted with a variety of Lewis bases, M, by the displacement reaction

$$\left(CH_3\right)_3 Si\,B^+ + M \rightarrow \left(CH_3\right)_3 Si\,M^+ + B \qquad (44)$$

which appears to be faster than the direct addition of $(CH_3)_3Si^+$ to M, a reaction which presumably requires a stabilizing collision for the adduct ion to be observed. When B = H_2O, proton transfer from $(CH_3)_3SiOH_2^+$ to M was observed in a number of cases.

The reaction of $(CH3)_3Si^+$ with alcohols, ROH, gives [M + $(CH_3)_3Si]^+$ adduct ions, $(CH_3)_3SiOH_2^+$ and R^+ ions.[90] The adduct ions were formed by displacement of water from the hydrated trimethylsilyl ion. It was possible to differentiate primary, secondary, and tertiary alcohols from the ratio of the abundance of the adduct ion to the hydrated trimethylsilyl ion. Distinctive reactions also are observed in the reaction of $(CH_3)_3Si^+$ with ethers.[91] The distinctive reactions of $(CH_3)_3Si^+$ with alcohols and ethers has led to the demonstration of the use of this reagent ion for the selective detection of oxygenated compounds in gasoline by GC/CIMS.[92]

G. Metal Ion Reagent Systems

There have been extensive studies of the reactions of metal ions with organic molecules,[93–95] although a relatively small number of these studies have been directed toward the analytical applications. A particularly useful approach has been the laser evaporation generation of metal ions in FTMS,[93,96] an approach which permits chemical ionization while still maintaining the low pressures necessary for effective operation of the instrument. Using this approach, Freiser and co-workers have investigated the reactions of Fe^+, Ti^+, Rh^+,

V^+, VO^+, Y^+, and La^+ with hydrocarbons[97–100] as well as Cu^+ and Fe^+ with oxygen-containing compounds.[101,102] Pattern recognition has been reported to be useful in the interpretation of these spectra for analytical applications.[103,104] More recently, it has been reported[105] that the reaction of laser-produced Mn^+ with chloroalkyl sulfides and organophosphonates has analytical utility while $InC_3H_6^+$ (produced by reaction of laser-produced In^+ with $(CH_3)_2CHCl$) has been suggested as a selective reagent for the ionization of alkenes and aromatics in the presence of alkanes.[106] Alkali metal ions may be produced by thermionic emission and some studies of the reactions of Li^+ and K^+ produced in this fashion have been reported.[107,108] Metal-containing ions also may be produced by EI of metal carbonyls, and the reaction of Co^+-containing ions[109,110] and Fe^+-containing ions[111–113] have been examined from an analytical viewpoint. In the latter work, it has been shown that collision-induced dissociation of the complex of Fe^+ with olefinic compounds is useful in determining the position of the double bond.[112,113]

H. Miscellaneous Positive Ion Reagent Systems

Keough[114] has examined in detail the use of dimethyl ether as a CI reagent gas. Under CI conditions (~0.2 torr), the major reactant ions are $CH_3OCH_2^+$ and $(CH_3)_2OH^+$ which undergo structure-specific ion/molecule reactions with the additive. These include protonation, hydride ion abstraction, and formation of $[M + CH_3OCH_2]^+$ adduct ions. Since relative response factors were very dependent on the structure of the molecule, selective ionization appears to be a possibility. Lange[115,116] has examined the use of ethylene oxide as a reagent gas for the analysis of alkanes, alkenes, alcohols, aldehydes and aminodienes. The major reactant ions are $C_2H_5O^+$, $C_2H_4O^+$, $C_2H_3O^+$ and $C_3H_5O^+$. The major product ions observed were $[M + H]^+$, $[M + C_2H_3O]^+$, and $[M + C_2H_5O]^+$. Jalonen and Oskman[117] have investigated the use of acetonitrile as a reagent gas for the chemical ionization of alcohols. The major reactant ion was the protonated nitrile and the main reaction channels were protonation, elimination of H_2O from MH^+, hydride ion abstraction, adduct ion formation, and substitution to form $RNCCH_3^+$ from ROH. Acrylonitrile at CI pressures forms mainly the protonated species CH_2CHCNH^+ and its reactions with alcohols, unsaturated acids and esters, and olefins and aromatic compounds have been reported briefly.[118] The major reactions are by protonation and adduct ion formation; the reactions with alcohols are similar to the reactions of protonated acetonitrile.

There have been a number of other reagent systems which have seen even less extensive study. A more complete discussion of these has been presented recently.[119] Of particular interest are reagent systems/ions designed to probe the stereochemistry or chirality of the sample; this topic will be addressed in Chapter 6.

III. NEGATIVE ION REAGENT SYSTEMS

A. Electron Capture Reagent Systems

Electron capture is not, strictly speaking, a chemical ionization process but

TABLE 6
Rate Constants for Electron Thermalization

Gas	k_e (cm^{3}s^{-1})
Ar	1.3×10^{-13}
He	6.4×10^{-12}
N_2	2.2×10^{-11}
H_2	1.1×10^{-10}
CH_4	8.6×10^{-10}
neo–C_5H_{12}	2.1×10^{-9}
CO_2	5.8×10^{-9}
NH_3	5.9×10^{-9}

Note: Data from Warman and Sauer.[121]

usually is included as such because a high pressure of a buffer or reagent gas is used. As discussed in Section IV.A, Chapter 2, the capture of electrons by polyatomic molecules is a resonance process which normally requires electrons of near-thermal energy. The buffer gas used in ECCI serves to "thermalize" the electrons emitted from the filament by inelastic scattering and dissociative ionization processes. The resultant quasi-thermal electrons are captured by the substrate molecule in either electron capture or dissociative electron capture processes. Ideally, the buffer gas should not form stable anions and buffer gases which have been used include H_2, He, CH_4, NH_3, N_2, Ar, CO_2 and i-C_4H_{10};[120] in all gases, only low yields of negative ions are obtained.

The sensitivity for ionization of a substrate molecule by electron capture might be expected to depend, in part, on the efficiency of the buffer gas in thermalizing the electrons. Table 6 presents second-order rate constants for the thermalization of low energy electrons.[121] Clearly, Ar, He, and N_2 are inefficient compared to polyatomic gases such as CH_4, CO_2, and NH_3. The datum for neo-C_5H_{12} is included as an approximation for i-butane. A further factor which may influence the sensitivity of ECCI is the efficiency of the buffer gas in collisionally stabilizing the M^{-*} formed in the electron capture process; in the absence of such stabilization, autodetachment of the electron will most likely occur. In general, polyatomic species such as CH_4, CO_2, i-C_4H_{10}, and NH_3 have higher efficiencies than monatomic and diatomic species in collisionally deactivating excited ions.[122,123] Recently, Sears and Grimsrud[124] have compared the relative sensitivities for detection of a number of nitrobenzene molecular anions in ECCI using a variety of buffer gases at the same pressure. With CO_2 and CH_4 buffer gases, similar sensitivities were obtained while with N_2 and Ar, the sensitivities were lower by about a factor of 30 and with He, the sensitivity was lower by a factor of ~100, in qualitative agreement with the above arguments. Xe showed a particularly low sensitivity, about a factor of 700 lower than CO_2 or CH_4.

A major attraction of ECCI, compared to other forms of CI or to EI, is the possibility of enhanced sensitivity. This arises because, as shown in Table 12, Chapter 2, electron capture rate constants as high as 10^{-7} cm^3 molecule^{-1} s^{-1}

TABLE 7
Relative Sensitivities for ECCI of Polycyclic Aromatic Hydrocarbons

| Compound | Relative sensitivity[a] | | EA(calc) (eV) |
	Oehme[125]	Brotherton and Gulick[126]	Hilpert[127]
Phenanthrene	0.5	6	0.03
Anthracene	6.8	3.1	0.49
Fluoranthene	1.1×10^2	13×10^2	0.63
Pyrene	0.4	10	0.45
Benz(a)anthracene	81	34	0.42
Benzo(b)fluoranthene	4.8×10^2	7.0×10^2	0.61
Benzo(k)fluoranthene	4.5×10^2	15×10^2	0.55
Benzo(e)pyrene	0.13	6.0	0.35
Benzo(a)pyrene	1.5×10^3	3.8×10^2	0.64

[a]   Relative to a sensitivity of 10 for EI of benzo(e)pyrene.

are possible compared to maximum rate constants for particle transfer ion/ molecule reactions in the range 2 to 4×10^{-9} cm^3 molecule^{-1} s^{-1}. Consequently, in favorable cases, ECCI may be much more sensitive than Brønsted acid or Brønsted base CI. Conversely, as the same table shows, electron capture rate constants can be as low as 10^{-13} cm^3 molecule^{-1} s^{-1} and ECCI does show variable and, sometimes, very low sensitivity. For example, Table 7 records relative sensitivities for detection of polycyclic aromatic hydrocarbons measured in two laboratories using CH$_4$ buffer gas.[125,126] Sensitivity variations of several hundred are observed, as well as distinct differences in the sensitivities measured in the two laboratories. A recent interlaboratory comparison of the limits of detection by ECCI also showed large differences even though conditions were standardized as far as possible;[128] these differences were attributed, in part, to different effective electron temperatures in the different instruments.

For the compounds of Table 7, at least, there appears to be a correlation of sensitivity with EA; only those compounds with an EA >0.5 eV show high responses. This points out the necessity of a substantial EA for the analyte before a strong electron capture ion signal will be observed. The presence of halogen or nitro groups (especially when attached to π-bonded carbon systems), highly conjugated carbon systems or a conjugated carbonyl system lead to a reasonably high EA. For compounds which do not have an appropriate EA, derivatization can be used to enhance the EA (and frequently, to enhance chromatographic properties). Perfluoroacyl, perfluorobenzoyl, and nitrobenzoyl or nitrobenzyl derivatives often are used although it is not always clear why a particular derivative was chosen. In contrast to electron capture gas chromatography, it is not sufficient that electron capture be facile, rather ions characteristic of the analyte should be formed since ions characteristic of the derivatizing agent will leave in doubt the identity of the analyte. As an example, Trainor and Vouros[129] have shown that perfluoroacyl derivatives of chlorophenols gave only ions characteristic of the acyl group and not the phenol. On the other hand, the pentafluorobenzyl ethers gave primarily phenoxide ions characteristic of the phenol.

Apart from variable sensitivity, a problem which appears to be particularly prevalent in ECCI is the nonreproducibility of the spectra observed; this makes the establishment of reliable data bases difficult. The ECCI mass spectra of many compounds (chlorine-containing compounds have been particularly well studied) depend rather strongly on experimental conditions such as source temperature, buffer gas pressure, buffer gas identity, impurities in the buffer gas, instrument used, instrument focusing conditions, and the amount of analyte introduced into the source. A few examples will suffice to illustrate these points. Stemmler and Hites[130] have recorded the CH_4 ECCI mass spectra of 361 environmental contaminants and related compounds at 100°C and 250°C source temperature; in many cases much more extensive fragmentation was observed at the higher temperature. The extent of formation of Cl^- in the ECCI mass spectra of polychlorinated phenoxyanisoles depends on the buffer gas used[131] while, for 1,2,3,4-tetrachlorodibenzo-*p*-dioxin, a dependence of Cl^- yield on the pressure of the buffer gas was observed in some cases but not in others.[132] Garland and Miwa[133] have observed that the CH_4 ECCI mass spectra of diazepam and nor-diazepam show primarily M^- at 100°C source temperature but exclusively Cl^- at source temperatures greater than 175°C. Addition of oxygen (1 part in 2000) to the methane led to predominant formation of $[M - H]^-$. An interlaboratory comparison[134] of the CH_4 ECCI mass spectra of 12 compounds in six laboratories, under operating conditions as standardized as possible, showed significant differences in relative ion intensities. The results obtained with the HP5895 quadrupole instrument in this and another study[135] showed lower abundances of low mass fragment ions than did the other instruments; this difference may be related to the potential applied to the ion focus lenses of the HP instrument.[136] Finally, Stemmler and Hites[137] have shown that the CH_4 ECCI mass spectra of a variety of hexachlorocyclopentadiene derivatives differed from those reported earlier[138] in that they observed mainly molecular anions while the earlier work reported mainly $[M + Cl]^-$ and other adduct ions. They showed that this difference arose from differences in sample partial pressures; high sample pressures favored adduct formation possibly through the reaction

$$M^{-\cdot} + M \rightarrow [M + Cl]^- + [M - Cl]^\cdot \qquad (45)$$

The rationalization of the remaining results is not straightforward. However, one should bear in mind that, for species of low EA, thermal electron detachment may be quite rapid at higher temperatures[139] and only ions of high EA (such as Cl^-) are likely to survive as anions. This implies that the sensitivity of ECCI may decrease with increasing temperature, although this has not been tested systematically.

In addition to the variations in ion yield of the products of electron capture reactions, ECCI mass spectra sometimes are complicated by the observation of ions which are not explainable on the basis of electron capture by the analyte.

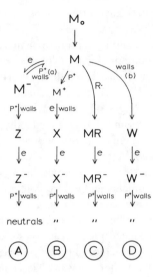

FIGURE 6. Pathways for forma-
tion of artifact peaks in ECCI. (From
Sears, L.J., Campbell, J.A., and
Grimsrud, E.P., *Biomed. Environ.
Mass Spectrom.*, 14, 401, 1987. With
permission.)

Some of these clearly arise from reactions of trace levels of ionic or neutral
impurities in the buffer gas. Low levels of H_2O or O_2 in the reagent gas will
lead to ions such as OH^-, O^-, and O_2^- which may react with the analyte;
conversely analyte ions may react with trace level neutral impurities. For
example, the $[M-19]^-$ ion sometimes observed in the ECCI mass spectra of
chlorinated compounds has been assigned as $[M + O - Cl]^-$ arising by reaction
of M^- with adventitious O_2.[140] In other cases, more complex origins are
indicated. Sears et al.[141] have considered in detail the ionization dynamics
within the high pressure electron capture source and have concluded that there
are four possible pathways for formation of unusual ions as outlined in Figure
6. Pathway A involves neutralization of the molecular anion at the walls or by
ion/ion recombination to form a species Z which captures an electron. Their
modelling indicated that this pathway could not provide an artifact signal
greater than that for M^-. Pathway B involves neutralization of positive ions to
form a species X which captures an electron. Their model calculations showed
that this route could not compete with a fast electron capture reaction of the
analyte but could result in a significant artifact signal if electron capture by M
was not rapid. They concluded that this pathway resulted in the formation of
the $[M - O]^-$ signals observed in the CH_4 ECCI mass spectra of a variety of
trifluroacetyl aminoanthracenes and aminophenanthrenes. Pathway C involves
the gas phase reaction of free radicals derived from the buffer gas with the
analyte followed by electron capture by MR. This route appears to be relatively
common when hydrocarbons are used as buffer gases; a number of examples
have been discussed by McEwen.[142] Pathway D is initiated by reaction of the

analyte on the walls of the source (or the chromatograph?) to produce a species W which undergoes electron capture. Model calculations showed that large artifact signals could arise from this route for analytes which do not undergo extremely fast electron capture. The $[M + 14]^-$ ion signal observed for fluorene has been attributed[143] to oxidation of the fluorene to fluorenone on the walls of the source with subsequent electron capture by the fluorenone. While these artifact signals may sometimes be a nuisance, they have been used under controlled conditions to differentiate polycyclic aromatic hydrocarbons.[144] Sears and Grimsrud[124] have reported that artifact signals are much reduced when CO_2 is used as the buffer gas compared to CH_4 as a buffer gas. However, work in this laboratory[145] has shown that artifact signals are observed using CO_2 buffer gas when the analyte is a poor electron capture species. These artifact signals are predominantly oxidation products such as $[M + O - 2H]^-$ for fluorene.

Clearly, the same alterations of the analyte can occur under other chemical ionization conditions. The point is that in ECCI the altered products may be ionized much more efficiently than the original analyte if the electron capture rate constants are appropriate. This preferential ionization of the artifact compounds is unlikely in other chemical ionization methods involving particle transfer since the rate constants show much less variation from compound to compound. Thus, in general, artifact peaks are much less prevalent in other CI methods. However, there are a number of cases where artifact peaks are observed in Brønsted acid and Brønsted base CI; some of these will be discussed in Chapter 5.

Finally, it might be noted that ECCI has often been called negative ion chemical ionization in the literature. As will be noted in the following section there are other negative ion chemical ionization methods; thus, this title is not sufficiently descriptive. The method also has been called electron capture negative ion mass spectrometry, a term which seems somewhat redundant.

B. Brønsted Base Reagent Systems

In negative ion chemical ionization mass spectrometry, Brønsted bases, B^-, often reacting by Reaction 46, play a role analogous to the role played by Brønsted acids in positive ion chemical ionization mass spectrometry. The thermochem-

$$B^- + M \rightarrow BH + [M - H]^- \qquad (46)$$

istry of Reaction 46 is given by $\Delta H = \Delta H_{acid}(M) - \Delta H_{acid}(BH) = PA([M - H]^-) - PA(B^-)$. As discussed in Section IV.D, Chapter 2, Reaction 46 is likely to be rapid for simple B^- provided it is exothermic. Thus, the reaction should be efficient provided BH is a weaker acid than M or, equivalently, provided $PA(B^-)$ is greater than $PA([M - H]^-)$. Since most of the exothermicity of the reaction is likely to reside in the B–H bond formed, little fragmentation of $[M - H]^-$ is expected. Table 8 lists the PAs for a number of anions which are candidate Brønsted base reagent ions. Also included in the table are the EAs of the neutral B, which are relevant to the tendency of the ion to react by charge exchange. The H^- ion does not appear to have been used as a CI reagent ion,

TABLE 8
Brønstead Base Reagent Systems

B⁻	PA(B⁻) (kcal mol⁻¹)	EA(B) (kcal mol⁻¹)
H⁻	400	17.4
NH₂⁻	404	18.0
OH⁻	391	42.2
O⁻	382	33.7
CH₃O⁻	381	36.2
F⁻	371	78.4
O₂⁻	353	10.1
Cl⁻	333	83.4
Br⁻	324	77.6

presumably because it is difficult to form it in good yield. The NH_2^- ion, produced by electron capture in NH_3 has seen only limited use.[146,147] The reactions of this species with a few esters have been reported;[147] [M – H]⁻ and carboxylate ions were the major products observed. The O⁻ and O_2^- ions react by several routes other than proton abstraction and will be discussed separately in later sections.

The OH⁻ ion has seen by far the greatest use as a Brønsted base reagent. It is produced by the following reaction sequence:

$$e_{therm} + N_2O \rightarrow O^- + N_2 \tag{47}$$

$$O^- + H_2/CH_4 \rightarrow OH^- + H^./CH_3^. \tag{48}$$

In early work,[148] $N_2O/He/H_2$ or $N_2O/He/CH_4$ mixtures in 1:1:1 ratios were used. It is more common now to use N_2O/CH_4 mixtures with the composition adjusted to eliminate the O⁻ signal. OH⁻ is a strong Brønsted base and will react with most organic molecules to give [M – H]⁻ ions; exceptions are alkyl amines and alkanes which are essentially unreactive. Alcohols give alkoxide ions frequently accompanied by loss of H_2 to give $[M - H-H_2]^-$ in low yield.[148–150] $[M - H-H_2O]^-$ ions frequently are observed in diols,[151,152] oxygenated terpenoids[150] and steroids.[153] Alkenes and alkylaromatic compounds give [M – H]⁻ ions which are reactive with N_2O and form $[M - H + N_2O]^-$ ([M + 43]⁻) and $[M-H+N_2O-H_2O]^-$ ([M + 25]⁻).[148,150] Esters, $RCOOR_1$, yield not only [M – H]⁻ but also carboxylate ions RCOO⁻;[148,150,153–155] elimination of R_1OH from the [M – H]⁻ ion also may be observed.[155] For methyl esters, it has been shown[156] that formation of carboxylate anions occurs not only by an S_N2 displacement reaction but also by attack of OH⁻ at the carbonyl carbon, Reaction 49.

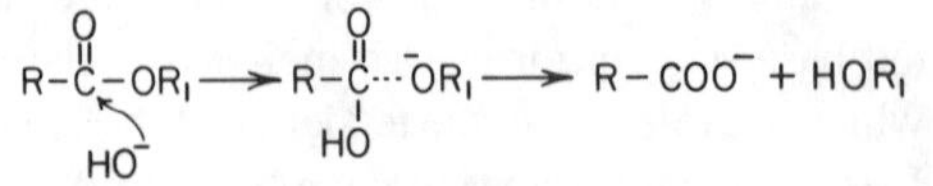

$$R-\overset{\overset{\displaystyle O}{\|}}{\underset{\underset{\displaystyle HO^-}{}}{C}}-OR_1 \longrightarrow R-\overset{\overset{\displaystyle O}{\|}}{\underset{\underset{\displaystyle HO}{}}{C}}\cdots\overset{-}{O}R_1 \longrightarrow R-COO^- + HOR_1 \tag{49}$$

For esters containing a β-hydrogen in R_1 carboxylate ions also can be produced by an elimination Reaction 50 illustrated for an ethyl ester.

$$HO^- \cdot H-CH_2-CH_2-O-CR \longrightarrow H_2O + C_2H_4 + RCOO^- \tag{50}$$

OH$^-$ chemical ionization also has been applied to a variety of more complex molecules such as glucoronides,[157] cardenolides,[158] trichothecenes,[159] and pyrrolizidine alkaloids.[160,161] [M − H]$^-$ ions were observed as well as a variety of fragment ions which were structurally informative.

The usefulness of CH$_3$O$^-$ as a CI reagent ion has not been investigated systematically. It is almost as strong a base as OH$^-$ and should react readily with many organic compounds by proton abstraction; the studies to date confirm this.[162] With their high EAs neither OH$^-$ or CH$_3$O$^-$ are likely to react by charge exchange. The CH$_3$O$^-$ ion is readily prepared by adding about 1% CH$_3$ONO to a moderating gas of CH$_4$. CH$_3$O$^-$ probably has little advantage over OH$^-$ as a Brønsted base reactant ion.

The chemical ionization reactions of F$^-$ with a variety of molecules have been investigated by Tiernan et al.[163] They found that electron impact on CHF$_3$ at 0.1 torr produced predominantly clusters of the type (HF)$_n$F$^-$ (n = 1 to 3) with practically no F$^-$, while electron impact on NF$_3$ at 0.1 torr produced largely F$^-$ with only a minor yield of (HF)F$^-$; the latter system was used for the studies they reported. It also has been reported[147] that electron impact on CF$_4$ produces primarily F$^-$. The F$^-$ ion is a weaker Brønsted base than OH$^-$ or CH$_3$O$^-$ and is not capable of abstracting a proton from alkanols, amines, alkanes, alkenes, or most nitriles. With the simple alkanols investigated, a low yield of [M + F]$^-$ was observed, however, the major product was (HF)F$^-$. With cycloalkane diols F$^-$ is reported to react by both proton abstraction to form [M − H]$^-$ and clustering to form [M + F]$^-$,[164] with the relative intensities depending on the stereochemistry of the diol. Amines gave only (HF)F$^-$, CN$^-$, and an unidentified product of m/z 42. Alkanes gave a very low yield of [M + F]$^-$ while alkenes and alkylbenzenes were unreactive. With the exception of acetonitrile, nitriles gave no [M − H]$^-$ ions, the major product being CN$^-$ presumably formed in a S$_N$2 displacement reaction. Esters were reported to give mainly the corresponding carboxylate anions and ethers gave alkoxide ions. More recent work[147] reported that acetate esters gave both [M − H]$^-$ and carboxylate ions as well as FCOCH$_2^-$, presumably formed by elimination of ROH from the tetrahedral adduct of F$^-$ with the ester CH$_3$COOR. Tiernan et al.,[163] reported that carboxylic acids, aldehydes, and ketones gave predominantly [M − H]$^-$ with low yields of [M + F]$^-$. F$^-$ has been used recently[165] for the selective ionization of naphthenic acids in crude oils and refinery wastewaters.

The Cl$^-$ ion is readily formed under CI conditions by dissociative electron capture by CH$_2$Cl$_2$, CHCl$_3$, CCl$_4$, or CF$_2$Cl$_2$.[120] If the reagent gas molecule contains a hydrogen as in CH$_2$Cl$_2$ and CHCl$_3$, moderately high yields of the adducts CH$_2$Cl$_3^-$ or CHCl$_4^-$ also will be formed. The Cl$^-$ ion is a weak Brønsted base and will abstract a proton only from very acidic compounds. In the absence of such acidic hydrogens, [M + Cl]$^-$ attachment ions frequently are formed. Thus, it has been reported[166] that carboxylic acids, amides, amino acids, aromatic amines and phenols give relatively intense attachment ions while aliphatic alcohols and amines, carbonyl compounds, ethers, nitroaromatics, and chlorinated hydrocarbons give lower yields of attachment ions and aliphatic and aromatic hydrocarbons, tertiary amines, and nitriles give no attachment ions under CI conditions. More recently, it has been reported[167] that dicarboxylic acids form $[M - H]^-$ ions while $[M - H]^-$ ions also have been reported[168] in the reaction of Cl$^-$ with a number of amino acids and other carboxylic acids. Simple amino acids form mainly $[M + Cl]^-$ while those with a second functionality capable of hydrogen bonding form mainly $[M - H]^-$.[169] Cl$^-$ chemical ionization is particularly useful for establishing molecular weights of complex organic molecules through formation of $[M + Cl]^-$ adduct ions;[170-173] in some cases, as in polysaccharides,[170] structural information also is obtained. The formation of the $[M + Cl]^-$ adduct ion has been used for the selective ionization of phenols in coal-derived liquids.[174] The reaction of Cl$^-$ with zinc(II) β-ketoenolates leads not only to formation of the adduct ion but also to replacement of the ligands by Cl$^-$.[175]

The Br$^-$ ion, which can be prepared by dissociative electron capture by CH_2Br_2, is an even weaker base than Cl$^-$ and reacts with substrates containing a modestly acidic hydrogen by formation of $[M + Br]^-$.[176] Surprisingly, with a number of dicarboxylic acids $[M - H]^-$ is the dominant product ion.[167]

C. O⁻· As a Reagent Ion

Electron capture by nitrous oxide (N_2O), using a non reactive buffer gas such as N_2 or a rare gas, is a convenient source of O$^-$.[177] The reactions of O$^-$ with a variety of organic molecules have been investigated using ICR techniques,[178-181] the flowing afterglow technique,[182] and medium pressure chemical ionization techniques.[177,183-187] A number of different types of reaction are observed with the relative importance of these reactions depending strongly on the structure of the substrate molecule. These reactions may be categorized as follows.

1. Hydrogen Atom Abstraction

$$O^- + M \rightarrow OH^- + [M - H]^{\cdot} \tag{51}$$

The OH$^-$ ion is one of the products of the reaction of O$^-$ with many organic molecules.[180,182] This OH$^-$ may subsequently react as a Brønsted base with the substrate to give $[M - H]^-$, consequently, O$^-$ CI mass spectra may depend on sample size.[180,186]

2. Proton Abstraction

$$O^- + M \rightarrow OH^{\cdot} + [M - H]^- \tag{52}$$

The O$^-$ ion is a relatively strong Brønsted base and is capable of abstracting a proton from many organic substrates, Frequently, $[M - H]^-$ is the most abundant ion in the CI mass spectrum; it may originate, in part, by reaction of OH^-, as discussed above.

3. H_2^+ Abstraction

$$O^{-\cdot} + M \rightarrow H_2O + [M - 2H]^{-\cdot} \qquad (53)$$

If the organic molecule contains suitably activated CH_2 groups H_2^+ abstraction may occur. The reaction was first observed with C_2H_4 where deuterium labeling showed that both hydrogens originated from the same carbon atom.[178] Subsequent work has shown that the reaction occurs with many 1-alkenes,[179] primary alkyl nitriles,[179,182] alkyl chlorides,[182] carbonyl compounds,[180,182,187] and a number of aromatic compounds.[177,186] Although the results for ethylene indicated that the two hydrogens came from the same carbon atom, this is not always the case. Thus, in H_2^+ abstraction from m-xylene, one H originates from each methyl group,[177] while in H_2^+ abstraction from propyne, the acetylenic hydrogen is involved to some extent.[188] In H_2^+ abstraction from ketones, the two hydrogens may originate from one alpha carbon or one hydrogen may originate from each alpha carbon.[187,189] For ketones at least, these $[M-2H]^{-\cdot}$ ions may undergo further fragmentation by Reactions 54 and 55.[187]

$$RCH_2C(=O)C^{-\cdot}CH_2R_1 \rightarrow R_1CH_2CCO^- + RCH_2 \qquad (54)$$

$$RCH^-C(=O)CH^{\cdot}CH_2R_1 \rightarrow RCH^-C(=O)CH = CH_2 + R_1^{\cdot} \qquad (55)$$

4. H Atom Displacement

$$O^{-\cdot} + M \rightarrow [M - H + O]^- + H^{\cdot} \qquad (56)$$

Almost all aromatic compounds investigated give a peak at $[M + 15]^-$ corresponding to the phenoxide ion derived by displacement of a hydrogen atom of the aromatic ring. Experiments with toluene-α, α, αd_3 show that an aromatic hydrogen is displaced exclusively.[190] Similarly, aldehydes react, in part, by H atom displacement to give the relevant carboxylate anion.[180,182,187]

5. Alkyl Group Displacement

$$O^{-\cdot} + M \rightarrow [M - R + O]^- + R^{\cdot} \qquad (57)$$

Displacement of an alkyl group, leading to the appropriate carboxylate anion, is observed in the reaction of $O^{-\cdot}$ with carbonyl compounds.[180,182,187] With symmetrical ketones only one carboxylate ion will be observed while for unsymmetrical ketones two carboxylate ions will be formed. Methyl-substituted aromatic compounds show low intensity $[M+1]^-$ ion signals arising by displacement of the methyl group by $O^{-\cdot}$.[177,186] The $O^{-\cdot}$ CI mass spectra of fluoroaromatic molecules also shows signals corresponding to displacement of the F by $O^{-\cdot}$.[177,190]

D. O_2^- As a Reagent Ion

A mixture of hydrogen and oxygen in an EI source has been used as a source of O_2^-,[162] while a Townsend discharge in pure O_2 at ~1 torr pressure also has been used to produce O_2^- as a reagent ion.[191-193] Negative ion CI mass spectra using CH_4/O_2 mixtures also have been measured;[194-196] the addition of O_2 to the methane was a measure adopted to improve the reproducibility of EC CI mass spectra which were affected by traces of adventitious oxygen in the system. Consequently, these spectra may reflect, in part, reactions of O_2^-; in addition, quasi-thermal electrons are likely to be present since, at these relatively low pressures, the lifetime with respect to autodetachment of the O_2^{-*} formed by electron capture is very short (Table 11, Chapter 2) and collisional stabilization is inefficient. At low pressures in a FTMS the reaction sequence

$$e_{therm} + N_2O \rightarrow O^- + N_2 \tag{58}$$

$$O^- + N_2O \rightarrow NO^- + NO \tag{59}$$

$$NO^- + O_2 \rightarrow O_2^- + NO \tag{60}$$

has been used[181] to produce O_2^-. At the much higher pressures of atmospheric pressure chemical ionization, collisional stabilization will be much more effective, and O_2^- and hydrated forms thereof are the major reactant ions using air[194,195] or N_2/O_2 mixtures[196] as carrier gases. The lifetimes of hot filaments are seriously reduced in contact with O_2; in addition, when rhenium filaments are used, background ions corresponding to the anions of rhenium trioxide and rhenium tetroxide are observed.[193] Discharge sources or atmospheric pressure ionization clearly are preferable when using O_2 as a reagent gas.

O_2^- is a stronger Brønsted base than Cl^- and will react by proton abstraction with a wider range of substrates. For example, it will abstract a proton from the three isomeric nitrophenols but not from the weaker acids, phenol and p-chlorophenol.[184] The EA of O_2^- is low, (~10 kcal mol^{-1}), and it also can react by charge exchange with molecules of suitably high EA.[162,191] Additional reaction modes are clustering to form $[M + O_2]^{-\cdot}$,[162,191] displacement of H to form $[M + O - H]^-$,[191] and displacement of Cl to form $[M + O - Cl]^-$.[195]

In many of these reactions, as well as the more complex reactions forming $C_6H_2Cl_2O_2^-$ in the O_2 CI of tetrachlorodibenzo-p-dioxins,[196,197] it is not clear whether the reaction is that of O_2^- with the substrate or reaction of M^- with neutral O_2. Probably both routes are possible, with the relative importance depending on experimental conditions.[198]

IV. SENSITIVITY OF CHEMICAL IONIZATION

A question of considerable importance is the sensitivity (ion current of additive per unit additive pressure) of the chemical ionization method. As will be seen below, the sensitivity depends strongly on the ion/molecule reaction rate constants discussed in Chapter 2. Absolute sensitivities are difficult to estimate because of unknown and variable ion source extraction efficiencies and mass spectrometer transmissions. However, the sensitivity of chemical ionization relative to EI on the same instrument can be derived. The following treatment for conventional CI and EI sources is based largely on the development by Field.[35]

The total ion current produced by EI is given by

$$I_{EI} = I_e QlN \tag{61}$$

where I_e is the ionizing electron current, Q is the molecular ionization cross section, l is the ionizing path length, and N is the concentration of molecules per cm^3. For CIMS, the additive ion current, I_{CI}, is given by the total number of reagent ions times the fraction $(1 - exp(-kNt))$ which react with the additive. Under CI conditions, the electron beam is completely attenuated, with the result that the total reagent ion current is given by the ionizing electron current, I_e, multiplied by the number, a, of ion pairs produced per electron. (Typically the energy required to produce an ion pair is ~30 eV[199] so that a $\approx$ 10 for 300 eV electrons). Thus,

$$I_{CI} = aI_e \left[1 - exp(-kNt)\right] \approx aI_e kNt \tag{62}$$

where k is the rate constant for reaction of the reagent ion(s) with the additive and t is the ion source reaction time available. For identical values of I_e and N we derive

$$I_{CI}/I_{EI} = akt/Ql \tag{63}$$

Substituting typical values of a = 10, k = 2×10^{-9} cm^3 molecule^{-1} s^{-1}, t = 10^{-5} s, l = 1 cm, and Q = 2×10^{-15} cm^2 molecule^{-1},[200] we obtain $I_{CI}/I_{EI} \approx 100$. This value represents the ratios of ion currents produced by equal electron beams in the ion source. Assuming the extraction efficiency from the ion source is proportional to the area of the ion exit slit, the collected ion current ratio can be reduced by a factor of 10 to 20 due to the use of a smaller ion exit slit in the CI source. In addition, the electron beam entrance slit often is reduced in a CI source, thus reducing I_e in CI sources relative to EI sources and further reducing the ratio. Nevertheless, these approximate calculations do indicate that the collected ion currents in CI are at least as intense as those collected in EI studies. In addition, the CI ion current may be concentrated in fewer ionic species.

Two extensions of this treatment deserve comment. As noted in Section IV.A, Chapter 2, electron capture reactions can have rate constants as high as 10^{-7} cm^3 molecule^{-1} s^{-1} compared to a maximum of ~4×10^{-9} cm^3 molecule^{-1} s^{-1} for ion/molecule reactions. Thus, in suitable cases, ECCI can have a sensitivity up to a factor of ~100 times greater than that for other CI methods. At the same time, Table 12, Chapter 2, shows that there is a great variation in electron capture rate constants with some being much lower than ion/molecule reaction rate constants. The high electron capture rate constants can be rationalized by noting that the collision rate constant for an ion and a neutral (Equation 14 or 18, Chapter 2) is proportional to $1/\mu^{1/2}$ there μ is the reduced mass of the colliding pair. The low mass of the electron makes μ small and, thus, increases the collisional rate constant.

In atmospheric pressure chemical ionization the reaction time, t, of Equation 63 will be greatly increased, particularly if ionization is effected by a ^{63}Ni source and the source is field-free; the reaction time enhancement is not as pronounced when a corona discharge is used for ionization because of the high field established between the needle and the plate containing the sampling orifice. These reaction time enhancements should lead to a greater sensitivity for APCI over that observed in medium pressure sources. Some of this enhancement is lost by the fact that lower ionizing currents are used, particularly in the ^{63}Ni source. Nevertheless, in favorable cases, APCI should be more sensitive than medium pressure CI. However, APCI using ambient air at room temperature is plagued by low sensitivity for detection of some substrates; indeed, sensitivities varying by a factor of 10^8 have been reported.[201] Under the operating conditions, the main protonating agent is $H_3O^+(H_2O)_h$ (h depending on the partial pressure of water in the ambient air) and the ionization of the substrate occurs by the reaction

$$H_3O^+\left(H_2O\right)_h + M \rightarrow MH^+\left(H_2O\right)_b + (h - b + 1)H_2O \qquad (64)$$

for which

$$\Delta G^\circ = GB\left(H_2O\right) - GB(M) - \Delta G^\circ_{0,h}\left(H_3O^+\right) + \Delta G^\circ_{0,b}\left(MH^+\right) \qquad (65)$$

where GB(X) is the gas phase basicity of X and the last two terms are the free energy changes for the respective solvation reactions

$$H_3O^+ + hH_2O \rightarrow H_3O^+\left(H_2O\right)_h \qquad (66)$$

$$MH^+ + bH_2O \rightarrow MH^+\left(H_2O\right)_b \qquad (67)$$

At room temperature and 5 torr H_2O, the most probable value of h is 5 or 6, $\Delta G^0_{0,h}(H_3O^+)$ is large and negative and overshadows the free energy gain by the (usually) weaker solvation of MH^+. Consequently, only for very strong bases (GB[M] > 200 kcal mol^{-1}, GB[H$_2$O] = 159 kcal mol^{-1}) will Reaction 64 have a negative ΔG^0. When ΔG^0 is negative, the reaction is under kinetic control[202,203] and substrates with such high GBs have uniformly high sensitivities. For most weaker bases, Reaction 64 is under thermodynamic control and the extent of ionization of the substrate (and hence, the sensitivity) is determined by the position of equilibrium. Several analytes, particularly carbon and sulfur bases, have very low sensitivities;[201] these species are distinguished by very low stabilities for the $MH^+(H_2O)_b$ hydrates. Sunner et al.[203] have shown that the sensitivities can be greatly enhanced for many substrates by operating the APCI instrument at elevated temperatures (up to 400°C). For example, the sensitivity for $(CH_3)_2S$ increased from 5×10^{-8} at 28°C to 0.11 at 430°C (relative to a sensitivity of 1 for pyridine).[203] The major effect of the temperature increase is to decrease the average degree of hydration and thus, to decrease the influence of the $\Delta G^0_{0,h}(H_3O^+)$ term in Equation 65.

REFERENCES

1. **Leventhal, J.J. and Friedman, L.**, Energy transfer in the de-excitation of $(H_3^+)^*$ by H_2, *J. Chem. Phys.*, 50, 2928, 1969.

2. **Bowers, M.T., Chesnavich, W.J., and Huntress, W.T.**, Deactivation of internally excited H_3^+ ions. Comparison of experimental product distributions of reactions of H_3^+ with CH_3NH_2, CH_3OH and CH_3SH with predictions of quasi-equilibrium theory calculations, *Int. J. Mass Spectrom. Ion Phys.*, 12, 357, 1973.

3. **Bowen, R.D. and Harrison, A.G.**, Chemical ionization mass spectra of selected C_3H_6O compounds, *Org. Mass Spectrom.*, 16, 159, 1981.

4. **Herman, J.A. and Harrison, A.G.**, Effect of reaction exothermicity on the proton transfer chemical ionization mass spectra of C_5 and C_6 alkanols, *Can. J. Chem.*, 59, 2125, 1981.

5. **Shannon, T.W. and Harrison, A.G.**, Concurrent ion-molecule reactions. II. Reactions in X-D_2 mixtures, *J. Chem. Phys.*, 43, 4206, 1965.

6. **Harrison, A.G. and Myher, J.J.**, Ion-molecule reactions in mixtures with D_2 or CD_4, *J. Chem. Phys.*, 46, 3276, 1967.

7. **Ikezoe, Y., Matsuoka, S., Takebe, M., and Viggiano, A.**, *Gas Phase Ion-Molecule Reaction Rate Constants Through 1986*, Maruzen Company, Tokyo, 1987.

8. **Munson, M.S.B. and Field, F.H.**, Chemical ionization mass spectrometry. I. General introduction, *J. Am. Chem. Soc.*, 88, 2621, 1966.

9. **Field, F.H. and Munson, M.S.B.**, Reactions of gaseous ions XIV. Mass spectrometric studies of methane at pressures up to 2 torr, *J. Am. Chem. Soc*, 87, 3289, 1965.

10. **Drabner, G., Poppe, A., and Budzikiewicz, H.**, The composition of the CH_4 plasma, *Int. J. Mass Spectrom. Ion Processes*, 97, 1, 1990.

11. **Brodbelt, J., Louris, J.N., and Cooks, R.G.**, Chemical ionization in an ion trap mass spectrometer, *Anal. Chem.*, 59, 1278, 1987.

12. **Munson, M.S.B., Franklin, J.L., and Field, F.H.**, High pressure mass spectrometric study of alkanes, *J. Phys. Chem.*, 68, 3098, 1964.

13. **Solka, B.H., Lau, A.Y.–K., and Harrison, A.G.**, Bimolecular reactions of trapped ions. VIII. Reactions in propane and methane-propane mixtures, *Can. J. Chem.*, 52, 1798, 1974.

14. **Shimizu, Y.**, Ions in propane chemical ionization mass spectrometry, *Bull. Nat. Res. Inst. Pollut. and Resour.* 15, 9, 1985.

15. **Field, F.H.**, Chemical ionization mass spectrometry. IX. Temperature and pressure studies with benzyl acetate and t-amyl acetate, *J. Am. Chem. Soc.*, 91, 2829, 1969.

16. **Hunt, D.F.**, Reagent gases for chemical ionization mass spectrometry, *Adv. Mass Spectrom.*, 6, 517, 1974.

17. **Rudewicz, P. and Munson, B.**, Effect of ammonia partial pressure on the sensitivities for oxygenated compounds in ammonia chemical ionization mass spectrometry, *Anal. Chem.*, 58, 2903, 1986.

18. **Hancock, R.A. and Hodges, M.G.**, A simple kinetic method for determining ion-source pressures for ammonia CIMS, *Int. J. Mass Spectrom. Ion Phys.*, 46, 329, 1983.

19. **Gupta, S.K., Jones, E.G., Harrison, A.G., and Myher, J.J.**, Reactions of thermal energy ions. VI. Hydrogen transfer ion-molecule reactions involving polar molecules, *Can. J. Chem.*, 45, 3107, 1967.

20. **Huntress, W.T. and Pinizzotto, R.F.**, Product distributions and rate constants for ion-molecule reactions in water, hydrogen sulfide, ammonia and methane, *J. Chem. Phys.*, 59, 4742, 1973.

21. **French, J.B., Davidson, W.R., Reid, N.M., and Buckley, J.A.**, Trace monitoring by tandem mass spectrometry, in *Tandem Mass Spectrometry*, McLafferty, F.W., Ed., John Wiley & Sons, New York, 1983.

22. **Good, A., Durden, D.A., and Kebarle, P.**, Ion-molecule reactions in pure nitrogen and nitrogen containing traces of water at total pressures of 0.5–4.0 Torr. Kinetics of clustering reactions forming $H^+(H_2O)_n$, *J. Chem. Phys.*, 52, 212, 1970.

23. **Good, A., Durden, D.A., and Kebarle, P.**, Mechanism and rate constants of ion-molecule reactions leading to formation of $H^+(H_2O)_n$ in moist oxygen and air, *J. Chem. Phys.*, 52, 222, 1970.

24. **Keough, T. and DeStefano, A.J.**, Factors affecting reactivity in ammonia chemical ionization mass spectrometry, *Org. Mass Spectrom.*, 16, 537, 1981.

25. **Cody, R. B.**, Separation of the reagent ions from the reagent gas in ammonia chemical ionization mass spectrometry, *Anal. Chem.*, 61, 2511, 1989.

26. **Westmore, J.B. and Alauddin, M.M.**, Ammonia chemical ionization mass spectrometry, *Mass Spectrom. Rev.*, 5, 381, 1986.

27. **van der Hart, W.J., Luijten, W.C.M.M., and van Thuijl, J.**, Ion/molecule reactions of ammonia and methylamine with chloro- and nitrobenzene. A comparison of chemical ionization and ion cyclotron resonance experiments, *Org. Mass Spectrom.*, 15, 463, 1980.

28. **Tholman, D. and Grützmacher, H.–Fr.**, FT-ICR study of the reaction of externally generated ammonia radical cations with chlorobenzene, *Org. Mass Spectrom.*, 24, 439, 1989.

29. **Stone, J.A., Splinter, D.E., McLaurin, J., and Wojtyniak, C.M.**, The formation of the anilinium ion from the ammonium ion in the chemical ionization of substituted benzenes, *Org. Mass Spectrom.*, 19, 375, 1984.

30. **Barry, R. and Munson, B.**, Selective reagents in chemical ionization mass spectrometry: diisopropyl ether, *Anal. Chem.*, 59, 466, 1987.

31. **Mollova, N.N., Handjieva, N.V., and Popov, S.S.**, Chemical ionization mass spectrometry with amines as reagent gases, VII. Amine chemical ionization mass spectrometry of some iridoid and secoiridoid glucosides, *Org. Mass Spectrom.*, 24, 1001, 1989: and references cited therein.

32. **Lindholm, E.**, Mass spectra and appearance potentials studied by use of charge exchange in a tandem mass spectrometer, in *Ion-Molecule Reactions*, Vol. 2, Franklin, J.L., Ed., Plenum Press, New York, 1972.

33. **Herman, J.A. and Harrison, A.G.**, Effect of protonation exothermicity on the chemical ionization mass spectra of some alkyl benzenes, *Org. Mass Spectrom.*, 16, 423, 1981.

34. **Harrison, A.G. and Kallury, R.K.M.R.**, Stereochemical applications of mass spectrometry. I. The utility of EI and CI mass spectrometry in differentiation of isomeric benzoin oximes and phenylhydrazones, *Org. Mass Spectrom.*, 15, 249, 1980.

35. **Field, F.H.**, Chemical ionization mass spectrometry, in *Mass Spectrometry, Series 1, Vol. 5, MTP Review of Science, Physical Chemistry*, Maccoll, A., Ed., Butterworths, London, 1972.

36. **Jardine, I. and Fenselau, C.**, Proton localization in chemical ionization fragmentation, *J. Am. Chem. Soc.*, 98, 5086, 1976.

37. **Harrison, A.G. and Onuska, F.I.**, Fragmentation in chemical ionization mass spectrometry and the proton affinity of the departing neutral, *Org. Mass Spectrom.*, 13, 35, 1978.

38. **Milne, G.W.A., Axenrod, T., and Fales, H.M.**, Chemical ionization mass spectrometry of complex molecules. IV. Amino acids, *J. Am. Chem. Soc.*, 92, 5170, 1970.

39.	**Tsang, C.W. and Harrison, A.G.**, The chemical ionization of amino acids, *J. Am. Chem. Soc.*, 98, 1301, 1976.

40.	**Mason, R., Milton, D., and Harris, F.**, Proton transfer to the fluorine atom in fluorobenzene: temperature dependence in the gas phase, *J. Chem. Soc. Chem. Commun.*, 1453, 1987.

41.	**Nakata, H., Suzuki, Y., Shibata, M., Takahashi, K., Konishi, H., Takeda, N., and Tatematsu, A.**, Chemical ionization mass spectrometry of bifunctional compounds. The behaviour of bifunctional compounds on protonation, *Org. Mass Spectrom.*, 25, 649, 1990.

42.	**Herman, J.A. and Harrison, A.G.**, Structural and energetics effects in the fragmentation of protonated esters in the gas phase, *Can. J. Chem.*, 59, 2133, 1981

43.	**Bowen, R.D. and Harrison, A.G.**, Chemical ionization mass spectra of some C_4H_8O isomers, *J. Chem. Soc. Perkin Trans. 2*, 1544, 1981.

44.	**Tsang, C.W. and Harrison, A.G.**, The origin of mass spectra, in *Biochemical Applications of Mass Spectrometry*, Waller, G.R., Ed., John Wiley & Sons, New York, 1972.

45.	**Howe, I., Williams, D.H., and Bowen, R.D.**, *Mass Spectrometry: Principles and Applications*, McGraw-Hill, New York, 1981.

46.	**McAdoo, D.J.**, Ion-neutral complexes in unimolecular decompositions, *Mass Spectrom. Rev.*, 7, 363, 1988.

47.	**Leung, H.–W. and Harrison, A.G.**, Structural and energetics effects in the chemical ionization of halogen-substituted benzenes and toluenes, *Can. J. Chem.*, 54, 3439, 1976.

48.	**Leung, H.–W. and Harrison, A.G.**, Specific substituent effects in the dehalogenation of halobenzene derivatives by the gaseous Brønsted acid CH_5^+, *J. Am. Chem. Soc.*, 102, 1623, 1980.

49.	**Houriet, R., Parisod, G., and Gäumann, T.**, The mechanism of chemical ionization of n-paraffins, *J. Am. Chem. Soc.*, 99, 3599, 1977.

50.	**Rudat, M.A.**, Chemical ionization with $C F_4$, 29th Annual Conference on Mass Spectrometry and Allied Topics, Minneapolis, May 24–29, 1981.

51.	**Herman, J.A., Li, Y.–H., and Harrison, A.G.**, Energy dependence of the fragmentation of some $C_6H_{12}^+$ ions, *Org. Mass Spectrom.*, 17, 143, 1982.

52.	**Kebarle, P.**, Higher order reactions — ion clusters and ion solvation, in *Ion-Molecule Reactions*, Franklin, J.L., Ed., Plenum Press, New York, 1972.

53.	**Sieck, L.W.**, Determination of molecular weight distribution of aromatic components in petroleum products by chemical ionization mass spectrometry with chlorobenzene as reagent gas, *Anal. Chem.*, 55, 38, 1983.

54.	**Allgood, C., Lin, Y., Ma, Y.–C., and Munson, B.**, Benzene as a selective chemical ionization reagent gas, *Org. Mass Spectrom.*, 25, 497, 1990.

55.	**Harrison, A.G. and Lin, M.S.**, Stereochemical applications of mass spectrometry III. Energy dependence of the fragmentation of stereoisomeric methylcyclohexanols, *Org. Mass Spectrom.*, 19, 67, 1984.

56.	**Meot-Ner, M. and Field, F.H.**, Thermodynamic ionization energies and charge transfer reactions in benzene and substituted benzenes, *Chem. Phys. Lett.*, 44, 484, 1976.

57.	**Meot-Ner, M., Hamlet, P., and Field, F.H.**, Bonding energies in association ions of aromatic compounds. Correlations with ionization energies, *J. Am. Chem. Soc.*, 100, 5466, 1978.

58.	**Li, Y.–H., Herman, J.A. and Harrison, A.G.**, Charge exchange mass spectra of some C_5H_{10} isomers, *Can. J. Chem.*, 59, 1753, 1981.

59.	**Harrison, A.G.**, Charge exchange mass spectrometry, in *Ionic Processes in the Gas Phase*, Almoster Ferreira, M. A., Ed., Reidel, Dordrecht, 1984.

60.	**Keough, T., Mihelich, E.D. and Eickhoff, D.J.**, Differentiation of monoepoxide isomers of polyunsaturated fatty acids and fatty acid esters by low-energy charge exchange mass spectrometry, *Anal. Chem.*, 56, 1849, 1984.

61.	**Abbatt, J.A. and Harrison, A.G.**, Low-energy mass spectra of some aliphatic ketones, *Org. Mass Spectrom.*, 21, 557, 1986.

62.	**Maccoll, A.**, Low energy, low temperature mass spectra 6. A synoptic view, *Org. Mass Spectrom.*, 21, 601, 1986.

63. **Subba Rao, S.C. and Fenselau, C.**, Evaluation of benzene as a charge exchange reagent, *Anal. Chem.*, 50, 511, 1978.

64. **Sunner, J. and Szabo, I.**, The analytical use of chemical ionization in a tandem mass spectrometer, *Adv. Mass Spectrom.*, 7, 1383, 1978.

65. **Thomson, B.A., Sakuma, T., Fulford, J., Lane, D.A., Reid, N.M., and French, J.B.**, Fast *in situ* measurement of PCB levels in ambient air to ng m^{-3} levels using a mobile atmospheric pressure chemical ionization mass spectrometer system, *Adv. Mass Spectrom.*, 8, 1422, 1980.

66. **Ferrer-Correia, A.J., Jennings, K.R., and Sen Sharma, D.K.**, The use of ion-molecule reactions in the mass spectrometric location of double bonds, *Org. Mass Spectrom.*, 11, 867, 1976.

67. **Ferrer-Correia, A.J., Jennings, K.R., and Sen Sharma, D.K.**, The use of ion-molecule reactions in the mass spectrometric location of double bonds, *Adv. Mass Spectrom.*, 7, 287, 1978.

68. **Greathead, R.J. and Jennings, K.R.**, The location of double bonds in mono- and di-unsaturated compounds, *Org. Mass Spectrom.*, 15, 431, 1980.

69. **Chai, R. and Harrison, A.G.**, Location of double bonds by chemical ionization mass spectrometry, *Anal. Chem.*, 53, 34, 1981.

70. **Budzikiewicz, H.**, Structure elucidation by ion-molecule reactions in the gas phase: the location of C,C double and triple bonds, *Fresenius Z. Anal. Chem.*, 321, 150, 1985.

71. **Lankelma, J., Ayanoglu, E., and Djerassi, C.**, Double-bond location in long-chain polyunsaturated acids by chemical ionization mass spectrometry, *Lipids*, 18, 853, 1983.

72. **Hunt, D.F. and Ryan, J.F.**, Chemical ionization mass spectrometry studies: nitric oxide as a reagent gas, *J. Chem. Soc. Chem. Commun.*, 620, 1972.

73. **Lias, S.G., Bartmess, J.E., Leibman, J.F., Holmes, J.L., Levin, R.D., and Miallard, W.G.**, Gas-phase ion and neutral thermochemistry, *J. Phys. Chem. Ref. Data*, 17, (Suppl. 1), 1988.

74. **Daishima, S., Iida, Y., and Kanda, F.**, Evaluation of nitric oxide chemical ionization mass spectra of substituted benzenes, *Org. Mass Spectrom.*, 26, 486, 1991.

75. **Einolf, N. and Munson, B.**, High pressure charge exchange mass spectrometry, *Int. J. Mass Spectrom. Ion Phys.*, 9, 141, 1972.

76. **Chai, R. and Harrison, A.G.**, Mixtures of nitric oxide with hydrogen and with methane as chemical ionization gases, *Anal. Chem.*, 55, 969, 1983.

77. **Budzikiewicz, H.**, Chemical ionization mass spectrometry with nitric oxide (NO) as reagent gas, in *Studies in Natural Products Chemistry, Vol. 2*, Atta-ur-Rahman, Ed., Elsevier, Amsterdam. 1988.

78. **Hunt, D.F. and Harvey, T.M.**, Nitric oxide chemical ionization mass spectra of olefins, *Anal. Chem.*, 47, 2136, 1975.

79. **Malosse, C. and Einhorn, J.**, Nitric oxide chemical ionization mass spectrometry of long-chain unsaturated alcohols, acetates and aldehydes, *Anal. Chem.*, 62, 287, 1990.

80. **Einhorn, J. and Malosse, C.**, Optimized production of the acylium diagnostic ions in the chemical ionization NO$^+$ mass spectra of long-chain monoolefins, *Org. Mass Spectrom.*, 25, 49, 1990.

81. **Schneider, B. and Budzikiewicz, H.**, Experiments on the formation of acylium from alkenic compounds following chemical ionization with NO$^+$, *Org. Mass Spectrom.*, 26, 498, 1991.

82. **Burt, J.A., Dunn, J.L., McEwan, M.J., Sutton, M.M., Roche, A.E., and Schiff, H.I.**, Some ion-molecule reactions of H$_3^+$ and the proton affinity of H$_2$, *J. Chem. Phys.*, 52, 6062, 1970.

83. **Vairamani, M.**, Nitromethane as a substitute for nitric oxide in chemical ionization, *Org. Mass Spectrom.*, 25, 271, 1990.

84. **Polley, Jr., C.W. and Munson, B.**, Nitrous oxide as a reagent gas for positive ion chemical ionization mass spectrometry, *Anal. Chem.*, 55, 754, 1983.

85. **Harvey, D.J., Horning, M.G., and Vouros, P.,** Vapour phase bimolecular reactions of alkyl siliconium and other metal ions in mass spectrometry, *Org. Mass Spectrom.*, 5, 599, 1971.

86. **Odiorne, T.J., Harvey, D.J., and Vouros, P.,** Chemical ionization mass spectrometry using tetramethylsilane, *J. Phys. Chem.*, 76, 3217, 1972.

87. **Odiorne, T.J., Harvey, D.J., and Vouros, P.,** Reactions of alkylsiliconium ions under chemical ionization conditions, *J. Org. Chem.*, 38, 4274, 1973.

88. **Stillwell, R.N., Carroll, D.I., Nowlin, J.G., and Horning, E.C.,** Formation of trimethylsilyl molecular adduct ions in desorption chemical ionization mass spectrometry of nonvolatile organic compounds, *Anal. Chem.*, 55, 1313, 1983.

89. **Clemens, D. and Munson, B.,** Selective reagents in chemical ionization mass spectrometry: trimethylsilyl adduct ions, *Anal. Chem.*, 57, 2022, 1985.

90. **Orlando, R., Strobel, F., Ridge, D.P., and Munson, B.,** Selective reagents in chemical ionization mass spectrometry: tetramethylsilane with aliphatic alcohols, *Org. Mass Spectrom.*, 22, 597, 1987.

91. **Orlando, R., Ridge, D.P., and Munson, B.,** Selective reagents in chemical ionization mass spectrometry: tetramethylsilane with ethers, *Org. Mass Spectrom.*, 23, 527, 1988.

92. **Orlando, R., and Munson, B,,** Trimethylsilyl ions for selective detection of oxygenated compounds in gasoline by gas chromatography/chemical ionization mass spectrometry, *Anal. Chem.*, 58, 2788, 1986.

93. **Freiser, B.S.,** Applications of laser ionization/Fourier transform mass spectrometry to the study of metal ions and their clusters in the gas phase, *Anal. Chim. Acta*, 178, 137, 1985.

94. **Allison, J.,** The gas-phase chemistry of transition-metal ions with organic molecules, *Prog. Inorg. Chem.*, 34, 627, 1986.

95. **Eller, K. and Schwarz, H.,** Organometallic chemistry in the gas phase, *Chem. Rev.*, 91, 1121, 1991.

96. **Cody, R.B., Burnier, R.C., Reents, W.D., Carlin, T.J., McRery, D.A., and Freiser, B.S.,** Laser ionization source for ion cyclotron resonance. Application to atomic metal ion chemistry, *Int. J. Mass Spectrom. Ion Phys.*, 33, 37, 1980.

97. **Byrd, G.D., Burnier, R.C., and Freiser, B.S.,** Gas-phase ion-molecule reactions of Fe^+ and Ti^+ with alkanes, *J. Am. Chem. Soc.*, 104, 3565, 1982.

98. **Byrd, G.D. and Freiser, B.S.,** Gas-phase reactions of Rh^+ with hydrocarbons, *J. Am. Chem. Soc.*, 104, 5944, 1982.

99. **Jackson, T.C., Carlin, J.C., and Freiser, B.S.,** Gas-phase reactions of V^+ and VO^+ with hydrocarbons using Fourier transform mass spectrometry, *J. Am. Chem. Soc.*, 108, 1120, 1986.

100. **Huang, Y., Wise, M.B., Jackson, D.B. and Freiser, B.S.,** Gas-phase reactions of yttrium and lanthanum ions with alkanes by Fourier transform mass spectrometry, *Organometallics*, 6, 346, 1987.

101. **Burnier, R.C., Byrd, G.D., and Freiser, B.S.,** Copper(I) chemical ionization mass spectrometric analysis of esters and ketones, *Anal. Chem.*, 52, 1641, 1980.

102. **Burnier, R.C., Byrd, G.D., and Freiser, B.S.,** Gas-phase reactions of Fe^+ with ketones and ethers, *J. Am. Chem. Soc.*, 103, 4360, 1981.

103. **Forbes, R.A., Tews, E.C., Freiser, B.S., Wise, M.B., and Perone, S.P.,** Application of pattern recognition to metal ion chemical ionization mass spectra, *Anal. Chem.*, 58, 684, 1986.

104. **Forbes, R.A., Tews, E.C., Huang, Y., Freiser, B.S., and Perone, S.P.,** Evaluation of laser desorbed transition-metal ions as analytical chemical ionization reagents by pattern recognition, *Anal. Chem.*, 59, 1937, 1987.

105. **Chou, C.–C. and Long, S.R.,** Chemical ionization Fourier transform mass spectrometry of chemical warfare agent simulants using laser-produced metal ions, *Appl. Opt.*, 29, 4981, 1990.

106. **Chowdhury, A.K., Cooper, J.R., and Wilkins, C.L.,** Indium-alkene complex ions as reagents for selective chemical ionization, *Anal. Chem.*, 61, 86, 1989.

107. **Hodges, R.V. and Beauchamp, J.L.,** Application of alkali metal ions in chemical ionization mass spectrometry, *Anal. Chem.,* 48, 825, 1976.
108. **Bombick, D., Pinkston, J.D., and Allison, J.,** Potassium ion chemical ionization and other uses of an alkali thermionic emitter in mass spectrometry, *Anal. Chem.,* 56, 396, 1984.
109. **Lombarski, M. and Allison, J.,** The gas-phase chemistry of metal and metal-containing ions with multifunctional organic molecules- investigating the utility of such ions as chemical ionization reagents, *Int. J. Mass Spectrom. Ion Phys.,* 49, 281, 1983.
110. **Lombarski, M. and Allison, J.,** Metal ion chemical ionization. Part II. The gas-phase chemistry of Co^+ with thiols, butanone, 3-butenone, carboxylic acids, substituted butanones and substituted carboxylic acids, *Int. J. Mass Spectrom. Ion Processes,* 65, 31, 1985.
111. **Peake, D.A., Gross, M.L., and Ridge, D.P.,** Mechanism of the reaction of gas-phase iron ions with neutral olefins, *J. Am. Chem. Soc.,* 106, 4307, 1984.
112. **Peake, D.A. and Gross, M.L.,** Iron(I) chemical ionization and tandem mass spectrometry for locating double bonds, *Anal. Chem.,* 57, 115, 1985.
113. **Peake, D.A., Huang, S.–K., and Gross, M.L.,** Iron(I) chemical ionization for analysis of alkene and alkyne mixtures by tandem mass spectrometry or gas chromatography/Fourier transform mass spectrometry, *Anal. Chem.,* 59, 1557, 1987.
114. **Keough, T.,** Dimethyl ether as a reagent gas for organic functional group determination by chemical ionization mass spectrometry, *Anal. Chem.,* 54, 2540, 1982.
115. **Lange, C.,** Ethylene oxide (oxiran) as a reagent gas for long chain compounds and organic functional group determination by chemical ionization mass spectrometry, *Org. Mass Spectrom.,* 21, 524, 1986.
116. **Lange, C.,** Ethylene oxide (oxiran) as a reagent gas for chemical ionization mass spectrometry: 2-conjugated dienic compounds, *Org. Mass Spectrom.,* 22, 55, 1987.
117. **Jalonen, J. and Oksman, P.,** New ionization reagents in chemical ionization. Ion/molecule reactions of $C_2H_4N^+$ with alcohols, *Spectrosc. Int. J.,* 5, 111, 1987.
118. **Srinivas, R., Vairamani, M., Vishanadha Rao, G.K., and Mirza, U.A.,** Acrylonitrile: a new chemical ionization reagent, *Org. Mass Spectrom.,* 24, 435, 1989.
119. **Vairamani, M., Mirza, U.A., and Srinivas, R.,** Unusual positive ion reagents in chemical ionization mass spectrometry, *Mass Spectrom. Rev.,* 9, 235, 1990.
120. **Budzikiewicz, H.,** Negative chemical ionization (NCI) of organic compounds, *Mass Spectrom. Rev.,* 5, 345, 1986.
121. **Warman, J. and Sauer, M.R. Jr.,** An investigation of electron thermalization in irradiated gases using CCl_4 as an electron energy probe, *J. Chem. Phys.,* 62, 1971, 1975.
122. **Miasek, P.G. and Harrison, A.G.,** Energy transfer in ion-molecule collisions: collisional deactivation of $[C_5H_9]^{+*}$, *J. Am. Chem. Soc.,* 97, 714, 1975.
123. **Ahmed, M.S. and Dunbar, R.C.,** Collisional relaxation of photoexcited bromobenzene ions by various neutral partners, *J. Am. Chem. Soc.,* 109, 3215, 1987.
124. **Sears, L.J. and Grimsrud, E.P.,** Elimination of unexpected ions in electron capture mass spectrometry using carbon dioxide buffer gas, *Anal. Chem.,* 61, 2523, 1989.
125. **Oehme, M.,** Determination of isomeric polycyclic aromatic hydrocarbons in air particulate matter by high-resolution gas chromatography/negative ion chemical ionization mass spectrometry, *Anal. Chem.,* 55, 2290, 1983.
126. **Brotherton, S.A. and Gulick, Jr., W.M.,** Positive- and negative-ion chemical ionization gas chromatography/mass spectrometry of polynuclear aromatic hydrocarbons, *Anal. Chim. Acta,* 186, 101, 1986.
127. **Hilpert, L.R.,** Determination of polycyclic aromatic hydrocarbons and alkylated polycyclic aromatic hydrocarbons in particulate extracts using negative ion chemical ionization mass spectrometry, *Biomed. Environ. Mass Spectrom.,* 14, 383, 1987.
128. **Arbogast, B., Budde, W.L., Deinzer, M., Dougherty, R.C., Eichelberger, J., Foltz, R.L., Grimm, C.C., Hites, R.A., Sakashita, C., and Stemmler, E.,** Interlaboratory comparison of limits of detection in negative chemical ionization mass spectrometry, *Org. Mass Spectrom.,* 25, 191, 1990.
129. **Trainor, T.M. and Vouros, P.,** Electron capture negative ion chemical ionization of derivatized chlorophenols and chloroanilines, *Anal. Chem.,* 59, 601, 1987.

130. **Stemmler, E.A. and Hites, R.A.**, *Electron Capture Negative Ion Mass Spectra of Environmental Contaminants and Related Compounds*, VCH Publishers, New York, 1988.

131. **Campbell, J.R., Griffen, D.A., and Deinzer, M.L.**, Electron capture negative ion and positive ion chemical ionization mass spectrometry of polychlorinated phenoxyanisoles, *Org. Mass Spectrom.*, 20, 122, 1985.

132. **Laramee, J.A., Arbogast, B.C., and Deinzer, M.L.**, Electron capture negative ion chemical ionization mass spectrometry of 1,2,3,4-tetrachlorodibenzo-*p*-dioxin, *Anal. Chem.*, 58, 2907, 1986.

133. **Garland, W.A. and Miwa, B.J.**, The [M–1]⁻ ion in the negative chemical ionization mass spectra of diazepam and nordiazepam, *Biomed. Mass Spectrom.*, 10, 126, 1983.

134. **Stemmler, E.A., Hites, R.A., Arbogast, B., Budde, W.L., Deinzer, M.L., Dougherty, R.C., Eichelberger, J.W., Foltz, R.L., Grimm, C., Grimsrud, E.P., Sakashita, C., and Sears, L.J.**, Interlaboratory comparison of methane electron negative ion mass spectra, *Anal. Chem.*, 60, 781, 1988.

135. **Oehme, M., Stöckl, D., and Knöppel, H.**, Comparison of the reproducibility of negative ion chemical ionization mass spectra obtained by different reagent gases on two commercial quadrupole instruments, *Anal. Chem.*, 58, 554, 1986.

136. **Stemmler, E.A. and Hites, R.A.**, A systematic study of instrumental factors affecting electron capture negative ion mass spectra, *Biomed. Env. Mass Spectrom.*, 15, 659, 1988.

137. **Stemmler, E.A. and Hites, R.A.**, Methane enhanced negative ion mass spectra of hexachlorocyclopentadiene derivatives, *Anal. Chem.*, 57, 684, 1985.

138. **Dougherty, R.C., Dalton, J., and Biros, F.J.**, Negative chemical ionization mass spectra of polycyclic chlorinated insecticides, *Org. Mass Spectrom.*, 6, 1171, 1972.

139. **Grimsrud, E.P., Chowdhury, S., and Kebarle, P.**, Thermal electron detachment rate constants. The electron detachment from azulene⁻ and the electron affinity of azulene. *J. Chem. Phys.*, 83, 3983, 1985.

140. **Grimsrud, E.P., Chowdhury, S., and Kebarle, P.**, Gas phase reactions of NO_2^- with nitrobenzenes and quinones. Electron transfer, clusters and formation of phenoxide and quinoxide negative ions. Use of NO_2 as a negative ion chemical ionization reagent, *Int. J. Mass Spectrom. Ion Processes*, 68, 57, 1986.

141. **Sears, L.J., Campbell, J.A., and Grimsrud, E.P.**, Ionization dynamics with the high pressure electron capture mass spectrometer; the unusual spectra of derivatized polycyclic aromatic amines and perchlorinated unsaturated hydrocarbons, *Biomed. Environ. Mass Spectrom.*, 14, 401, 1987.

142. **McEwen, C.N.**, Radicals in analytical mass spectrometry, *Mass Spectrom. Rev.*, 5, 52, 1986.

143. **Stemmler, E.A. and Buchanan, M.V.**, Negative ions generated by reactions with oxygen in the chemical ionization source. I. Characterization of gas-phase and wall-catalyzed reactions of fluorene, anthracene and fluoranthene, *Org. Mass Spectrom.*, 24, 94, 1989.

144. **Stemmler, E.A. and Buchanan, M.V.**, Negative ions generated by reactions with oxygen in the chemical ionization source. II. The use of wall-catalyzed reactions for the differentiation of polycyclic aromatic hydrocarbons and their methyl derivatives, *Org. Mass Spectrom.*, 24, 705, 1989.

145. **Molleken, M.E.**, *Carbon Dioxide as a Moderating Gas in Electron Capture Chemical Ionization*, M.Sc. thesis, University of Toronto, Toronto, 1990.

146. **Hunt, D.F. and Sethi, S.K.**, Gas-phase ion/molecule isotope-exchange reactions: methodology for counting hydrogen atoms in specific structural environments by chemical ionization mass spectrometry, *J. Am. Chem. Soc.*, 102, 6953, 1980.

147. **Grützmacher, H.–Fr. and Grotemeyer, B.**, Fragmentation reactions of some aliphatic esters in the NCI(F⁻) and NCI(NH₂⁻) mass spectra, *Org. Mass Spectrom.*, 19, 135, 1984.

148. **Smit, A.L.C. and Field, F.H.**, Gaseous anion chemistry. Formation and reaction of OH⁻. Reactions of anions with N_2O. OH⁻ negative chemical ionization, *J. Am. Chem. Soc.*, 99, 6471, 1977.

149. **Houriet, R., Stahl, D., and Winkler, F.J.**, Negative chemical ionization of alcohols, *Environ. Health Perspect.*, 36, 63, 1980.

150. **Bruins, A.P.**, Negative chemical ionization mass spectrometry in the determination of compounds in essential oils., *Anal. Chem.*, 51, 967, 1979.

151. **Gäumann, T., Stahl, D., and Tabet, J.C.**, The mechanism of the loss of H_2 and H_2O from 1,4-cyclohexanediol and related compounds, *Org. Mass Spectrom.*, 18, 203, 1983.

152. **Winkler, F.J. and Stahl, D.**, Stereochemical effects on anion mass spectra of cyclic diols. Negative chemical ionization, collisional activation and metastable ion spectra, *J. Am. Chem. Soc.*, 100, 6779, 1978.

153. **Lin, Y.Y. and Smith, L.L.**, Chemical ionization mass spectrometry of steroids and other lipids, *Mass Spectrom. Rev.*, 3, 319, 1984.

154. **Bambagiotti, M., Giannellini, V., Coran, S.A., Vincieri, F.F., Moneti, G., Selva, A., and Traldi, P.**, Negative ion chemical ionization and collisional activation mass spectrometry of naturally occurring bornyl C_4–C_5 esters, *Biomed. Mass Spectrom.*, 9, 495, 1982.

155. **Bambagiotti, M., Coran, S.A., Giannellini, V., Vincieri, F.F., Daolio, S., and Traldi, P.**, Hydroxyl ion negative chemical ionization and collisionally activated dissociation mass analyzed kinetic energy spectra for an easy mass spectrometric characterization of fatty acid methyl esters, *Org. Mass Spectrom.*, 18, 133, 1983.

156. **Riveros, J.M., José, S.M., and Takashima, K.**, Gas-phase nucleophilic displacement reactions, *Adv. Phys. Org. Chem.*, 21, 197, 1985.

157. **Bruins, A.P.**, Negative ion desorption chemical ionization mass spectrometry of some underivatized glucoronides, *Biomed. Mass Spectrom.*, 8, 31, 1981.

158. **Bruins, A.P.**, Negative ion desorption chemical ionization mass spectrometry of digitoxin and related cardenolides, *Int. J. Mass Spectrom. Ion Phys.*, 48, 185, 1983.

159. **Brumley, W.C., Andrzejewski, D.A., Trucksess, E.W., Dreifuss, D.A., Roach, J.A.G., Eppley, R.M., Thomas, F.J., Thorpe, C.W., and Sphon, J.A.**, Negative chemical ionization mass spectrometry of trichothecenes. Novel fragmentation under OH⁻ conditions, *Biomed. Mass Spectrom.*, 9, 451, 1982.

160. **Dreifuss, P.A., Brumley, W.C., Sphon, J.A., and Caress, E.A.**, Negative ion chemical ionization mass spectrometry of pyrrolizidine alkaloids with hydroxide reactant ion, *Anal. Chem.*, 55, 1036, 1983.

161. **Huizing, H.J., de Boer, F., Hendricks, H., Balraading, W., and Bruins, A.P.**, Positive and negative ion chemical ionization mass spectrometry of trimethylsilyl derivatives of pyrrolizidine alkaloids using NH_4^+ and OH⁻ as the reactant ions, *Biomed. Environ. Mass Spectrom.*, 13, 293, 1986.

162. **Hunt, D.F., Stafford, G.C., Crow, F.W., and Russell, J.W.**, Pulsed positive negative ion chemical ionization mass spectrometry, *Anal. Chem.*, 48, 2098, 1976.

163. **Tiernan, T.O., Chang, C., and Cheng, C.C.**, Formation and reactions of ions relevant to chemical ionization mass spectrometry. I. CI mass spectra of organic compounds produced by F⁻ reactions, *Env. Health Perspect.*, 36, 47, 1980.

164. **Winkler, F.J. and Stahl, D.**, Intramolecular ion solvation effects on gas-phase acidities and basicities, *J. Am. Chem. Soc.*, 101, 3685, 1979.

165. **Dzidic, I., Somerville, A.C., Raia,, J.R. and Hart, H.V.**, Determination of naphthenic acids in California crudes and refinery wastewaters by fluoride ion chemical ionization mass spectrometry, *Anal. Chem.*, 60, 1318, 1988.

166. **Tannenbaum, H.P., Roberts, J.D., and Dougherty, R.C.**, Negative chemical ionization mass spectrometry — chloride attachment spectra, *Anal. Chem.*, 47, 49, 1975.

167. **Vairamani, M. and Saraswathi, M.**, Negative ion chemical ionization (Br⁻) mass spectra of dicarboxylic acids, *Org. Mass Spectrom.*, 24, 355, 1989.

168. **Duarte, M.F.N., Hutchinson, D.W., and Jennings, K.R.**, Chemical ionization mass spectrometry of L-Dopa and L-trytophan and their derivatives in tumor samples, *Org. Mass Spectrom.*, 20, 476, 1985.

169. **Vairamani, M., Srinivas, R., and Viswanaha Rao, G.K.**, Negative ion chemical ionization (Cl⁻) mass spectra of amino acids, *Biomed. Environ. Mass Spectrom.*, 17, 299, 1988.

170. **Ganguly, A.K., Cappuccino, N.F., Fujiwara, H., and Bose, A.K.**, Convenient mass spectral technique for structural studies in oligosaccharides, *J. Chem. Soc. Chem. Commun.*, 148, 1979.

171. **Bose, A.K., Fujiwara, H., and Promanik, B.N.**, Negative chemical ionization mass spectra of multifunctional compounds of biological interest, *Tetrahedron. Lett.*, 4017, 1979.

172. **Madhusudanan, K.P., Gupta, S., and Bhakuni, D.S.**, Anion mass spectra of abnormal *Erythrina* alkaloids, *Indian. J. Chem.*, 22B, 907, 1983.

173. **Levsen, K., Schäfer, K.H., and Dobberstein, P.**, On-line liquid chromatographic/mass spectrometric studies using a moving belt interface and negative chemical ionization, *Biomed. Mass Spectrom.*, 11, 308, 1984.

174. **Anderson, G.B., Johns, R.B., Porter, Q.N., and Strachan, M.G.**, Freon-113 as a reagent gas in the negative chemical ionization of phenols and acidic fractions of coal-derived liquids, *Org. Mass Spectrom.*, 19, 583, 1984.

175. **Dillow, G.W. and Gregor, I.K.**, Halide ion and radical reactions of Zn(II) β-ketoenolates in the gas phase, *Inorg. Chim. Acta*, 86, L67, 1984.

176. **Caldwell, G.W., Masucci, J.A., and Ikonomou, M.G.**, Negative ion chemical ionization mass spectrometry — binding of molecules to bromide and iodide anions, *Org. Mass Spectrom.*, 24, 8, 1989.

177. **Bruins, A.P., Ferrer-Correia, A.J., Harrison, A.G., Jennings, K.R., and Mitchum, R.K.**, Negative ion chemical ionization mass spectra of some aromatic compounds using O⁻ as the reagent ion, *Adv. Mass Spectrom.*, 7, 355, 1978.

178. **Goode, G.C. and Jennings, K.R.**, Reactions of O⁻ with some unsaturated hydrocarbons, *Adv. Mass Spectrom.*, 6, 797, 1974.

179. **Dawson, J.H.J. and Jennings, K.R.**, Production of gas-phase radical anions by reaction of O⁻ with organic substrates, *J. Chem. Soc. Faraday Trans. II*, 72, 700, 1976.

180. **Harrison, A.G. and Jennings, K.R.**, Reactions of O⁻ with carbonyl compounds, *J. Chem. Soc. Faraday Trans I*, 72, 1601, 1976.

181. **Van Orden, S.L., Malcolmson, M.E., and Buckner, S.W.**, Mechanistic and kinetic aspects of chemical ionization mass spectrometry of polynuclear aromatic hydrocarbons and their halogen-substituted analogues using oxidizing reagents. A gas chromatographic-mass spectrometric and Fourier transform mass spectrometric study, *Anal. Chim. Acta*, 246, 199, 1991.

182. **Grabowski, J.J. and Melly, S.J.**, Formation of carbene radical anions; gas-phase reactions of the atomic oxygen anion with organic neutrals, *Int. J. Mass Spectrom. Ion Processes*, 81, 147, 1987.

183. **Jennings, K.R.**, Negative chemical ionization mass spectrometry, in *Mass Spectrometry, Specialist Periodical Report*, Vol. 4, Chemical Society, London, 1977.

184. **Jennings, K.R.**, Investigation of selective reagent ions in chemical ionization mass spectrometry, in *High Performance Mass Spectrometry*, Gross, M.L., Ed., Am. Chem. Soc., Washington, 1978.

185. **Jennings, K.R.**, Chemical ionization mass spectrometry, in *Gas Phase Ion Chemistry*, Bowers, M.T., Ed., Academic Press, New York, 1979.

186. **Harrison, A.G. and Tong, H.–Y.**, Characterization of C_8H_{10} alkylbenzenes by negative ion mass spectrometry, *Org. Mass Spectrom.*, 23, 135, 1988.

187. **Marshall, A., Tkaczyk, M., and Harrison, A.G.**, O⁻ chemical ionization of carbonyl compounds, *J. Am. Soc. Mass Spectrom.*, 2, 292, 1991.

188. **Dawson, J.H.J., Kaandorp, T.A.M. and Nibbering, N.M.M.**, A gas phase study of the ions $C_3H_2^-$ and $C_3H_3^-$ generated from the reaction of O⁻ with propyne, *Org. Mass Spectrom.*, 12, 330, 1977.

189. **Dawson, J.H.J., Noest, A.J., and Nibbering, N.M.M.,**, 1,1- and 1,3-elimination of water from the reaction complex of O⁻ with 1,1,1-trideuteroacetone, *Int. J. Mass Spectrom. Ion Phys.*, 30, 189, 1979.

190. **Merrett, K., Young, A.B., and Harrison, A.G.**, The O⁻ chemical ionization of aromatic and fluoroaromatic compounds, to be published.

191. **Hunt, D.F., McEwen, C.N., and Harvey, T.M.**, Positive and negative ion chemical ionization mass spectrometry using a Townsend discharge ion source, *Anal. Chem.*, 47, 1730, 1975.

192. **Miles, W.F., Gurprasad, N.P., and Malis, G.P.**, Isomer-specific determination of hexachlorodibenzo-*p*-dioxins by oxygen negative chemical ionization mass spectrometry gas chromatography and high pressure liquid chromatography, *Anal. Chem.*, 57, 1133, 1985.

193. **Miles, W.F. and Gurprasad, N.P.**, Oxygen negative ion chemical ionization mass spectrometry of trichothecenes, *Biomed. Mass Spectrom.*, 12, 653, 1985.

194. **Lane, D.A., Thomson, B.A., Lovett, A.M., and Reid, N.M.**, Real time tracking of industrial emissions through populated areas using a mobile APCI mass spectrometer system, *Adv. Mass Spectrom.*, 8, 1480, 1980.

195. **Dzidic, I., Carroll, D.I., Stillwell, R.N., and Horning, E.C.**, Atmospheric pressure ionization (API) mass spectrometry; formation of phenoxide ions from chlorinated aromatic compounds, *Anal. Chem.*, 47, 1308, 1975.

196. **Mitchum, R.K., Althaus, J.R., Korfmacher, W.A., and Moler, G.F.**, Application of negative ion atmospheric pressure ionization (NIAPI) mass spectrometry for trace analysis, *Adv. Mass Spectrom.*, 8, 1415, 1980.

197. **Hunt, D.F., Harvey, T.M., and Russell, J.W.**, Oxygen as a reagent gas for the analysis of 2,3,7,8-tetrachlorodibenzo-p-dioxin. Negative ion chemical ionization mass spectrometry. *J. Chem. Soc. Chem. Commun.*, 151, 1975.

198. **Mitchum, R.K., Korfmacher, W.A., and Althaus, J.R.**, Mechanism for semiquinone formation in the oxygen negative chemical ionization mass spectrometry of 2,3,7,8-tetrachlorodibenzo-p-dioxin, *Org. Mass Spectrom.*, 19, 63, 1984.

199. **Spinks, J.W.T. and Woods, R.J.**, *An Introduction to Radiation Chemistry*, John Wiley & Sons, New York, 1964.

200. **Harrison, A.G., Jones, E.G., Gupta, S.K., and Nagy, G.P.**, Total cross sections for ionization by electron impact, *Can. J. Chem.*, 44, 1967, 1966.

201. **Sunner, J., Nicol, G., and Kebarle, P.**, Factors determining relative sensitivity of analytes in positive mode atmospheric pressure ionization mass spectrometry, *Anal. Chem.*, 60, 1300, 1988.

202. **Nicol, G., Sunner, J., and Kebarle, P.**, Kinetics and thermodynamics of protonation reactions; $H_3O^+(H_2O)_h + B \rightarrow BH^+(H_2O)_b + (h - b + 1)H_2O$ where B is a nitrogen, oxygen or carbon base, *Int. J. Mass Spectrom. Ion Phys.*, 84, 135, 1988.

203. **Sunner, J., Ikonomou, M.G., and Kebarle, P.**, Sensitivity enhancements obtained at high temperatures in atmospheric pressure ionization mass spectrometry, *Anal. Chem.*, 60, 1308, 1988.

Chapter 5

CHEMICAL IONIZATION MASS SPECTRA

I. INTRODUCTION

An astonishing number of chemical compounds have been investigated by chemical ionization methods. It is neither possible nor practical to review all of these studies. The present review will concentrate on those classes of compounds for which systematic or substantial studies have been carried out, thus permitting general conclusions to be drawn. For the most part, the discussion will be limited to monofunctional compounds to illustrate the role of the functionality in determining the chemical ionization mass spectrum.

II. ALKANES (C_nH_{2n+2})

The CH_4 CI mass spectra of an extensive series of linear and branched alkanes have been reported by Field et al.[1] The spectra of the n-alkanes are characterized by abundant $[M - H]^+$ ions which account for 25 to 45% of the total additive ionization, as shown by the mass spectrum of n-decane presented as the first entry in Table 1. The $[M - H]^+$ ions originate by nominal hydride ion abstraction by both CH_5^+ (Reaction 1) and $C_2H_5^+$ (Reaction 2). The CI mass spectra also show lower mass alkyl ions which originate both by fragmentation of the $[M - H]^+$ ion (Reaction 3) and by nominal alkide ion abstraction reactions (Reaction 4). Detailed studies, using ICR techniques, have been made[2,3]

$$CH_5^+ + n - C_6H_{14} \rightarrow C_6H_{13}^+ + H_2CH_4 \tag{1}$$

$$C_2H_5^+ + n - C_6H_{14} \rightarrow C_6H_{13}^+ + C_2H_6 \tag{2}$$

$$C_6H_{13}^+ \rightarrow C_4H_9^+ + C_2H_4$$

$$\rightarrow C_3H_7^+ + C_3H_6 \tag{3}$$

$$CH_5^+ + n - C_6H_{14} \rightarrow C_5H_{11}^+ + CH_4 + CH_4$$

$$\rightarrow C_4H_9^+ + C_2H_6 + CH_4 \tag{4}$$

$$\rightarrow C_3H_7^+ + C_3H_8 + CH_4$$

113

TABLE 1
CH$_4$ Chemical Ionization Mass Spectra of Alkanes

Alkane	Alkyl ion						
	C$_4$	C$_5$	C$_6$	C$_7$	C$_8$	C$_9$	C$_{10}$
n-C$_{10}$	0.14	0.22	0.22	0.10	0.01	0.005	0.31
2-Me-5-Et-C7	0.11	0.21	0.14	0.04	0.05	0.08	0.20
2,3,6-Me$_3$-C$_7$	0.17	0.26	0.19	0.03	0.004	0.10	0.09
2,2,4-Me$_3$-C$_7$	0.32	0.19	0.17	0.04	0.004	0.12	0.06
2,2,3,3-Me$_4$-C$_6$	0.28	0.27	0.16	0.12	0.001	0.09	0.005

Note: Numbers are fraction of total ionization.

of the reactions of CH$_5^+$ and C$_2$H$_5^+$ with model alkanes. These studies have shown that C$_2$H$_5^+$ reacts entirely by hydride ion abstraction (Reaction 2) while CH$_5^+$ reacts by both Reactions 1 and 4. These two reactions may be viewed either as hydride and alkide abstraction reactions or as protonation followed by fragmentation. Evidence for two forms of protonated ethane, one protonated at the C–C bond and one protonated at the C–H bond have been presented;[4] for the higher alkanes these protonated species are thermochemically unstable with respect to fragmentation to alkyl ions. Reactions 1 and 2 could be pictured as the net result of protonation at a C–H bond while the nominal alkide ion abstraction Reaction 4 could equally well be pictured as the net result of protonation at the C–C bonds of the alkane. No matter how it is formed, the [M – H]$^+$ ion is known to undergo fragmentation by olefin elimination to form lower alkyl ions.[5] The low energy modes of fragmentation of alkyl ions have been predicted from thermochemical arguments by Bowen and Williams.[6]

The [M – H]$^+$ ion abundance for the n-alkanes is considerably greater than the ions characteristic of molecular mass (M$^+$, [M – H]$^+$) are in the electron ionization mass spectrum, although in the CI mass spectrum the [M – H]$^+$ intensity decreases rapidly with increasing source temperature.[5] In addition, with increased branching of the alkane the [M – H]$^+$ intensity drops markedly, as the spectra in Table 1 show; thus, for 2,2,3,3-tetramethylhexane the [M – H]$^+$ ion signal accounts for only 0.5% of the total additive ionization. This decrease in the [M – H]$^+$ intensity with increased branching can be attributed to an increased tendency for CH$_5^+$ to attack the C–C bonds (Reaction 4) at the point of chain branching and to a more facile fragmentation of the [M – H]$^+$ alkyl ion, which can readily form stable secondary or tertiary alkyl ions by olefin elimination (Reaction 3).

In an attempt to obtain greater abundances of ions characteristic of the molecular mass, Hunt and co-workers[7,8] have determined the CI mass spectra of a variety of alkanes using NO$^+$ as the reagent ion. A selection of their results is presented in Table 2. The major ion observed in all spectra is the [M – H]$^+$ ion, even for branched alkanes. Thus, the NO CI mass spectra are to be preferred for providing molecular mass information for alkanes. A drawback to the use of NO is the formation of [M–3H]$^+$ and [M–2H + NO]$^+$ ions with many of the alkanes; ions of the same m/z ratio are formed in the NO CI of the corresponding alkene. In the alkanes, these ions probably arise by oxidation of the alkane to an alkene

TABLE 2
NO Chemical Ionization Mass Spectra of Alkanes

Alkane	[M – H]⁺	[M–3H]⁺	[M–2H + NO]⁺	Alkyl ion			
				C_3	C_4	C_5	C_6
n-$C_{10}H_{22}$	80.0	2.4	1.6	2.4	2.4	1.6	9.6
n-$C_{34}H_{70}$	71.4	14.3	12.9	—	—	—	—
2,2-$Me_2C_4H_8$	86.2	—	10.3	1.7	0.9	0.9	—
2,2,4-$Me_3C_5H_{10}$	64.5	—	7.7	2.6	16.1	3.9	—
2,4,6-$Me_3C_7H_{14}$	74.1	0.7	0.1	—	3.4	10.4	11.1

Note: Numbers are % of total ionization

with subsequent reaction of the alkene to give the [M – H]⁺ and [M + NO]⁺ ions of the alkene (see the following section). The extent of oxidation is less for NO diluted with N_2 than for pure NO as the reagent gas,[7] and oxidation products were not observed when NO/H_2 mixtures were used to produce NO⁺.[9]

The EAs of alkanes are negative as are the EAs of many alkyl radicals;[10] consequently, alkanes are essentially unreactive under negative ion CI conditions.

III. ALKENES AND CYCLOALKANES (C_nH_{2n})

The CH_4 CI mass spectra of a variety of alkenes have been reported by Field.[11] The spectra of mono-olefins consist of two series of ions, the $[C_nH_{2n+1}]^+$ alkyl ions consisting of the MH⁺ ion and fragments derived therefrom and $[C_nH_{2n-1}]^+$ alkenyl ions consisting of the [M – H]⁺ ion and lower mass fragment ions derived therefrom. Typical spectra are shown in Figure 1 using the isomeric *n*-decenes as examples. In forming the [M – H]⁺ ions, allylic hydrogens apparently are most readily abstracted since the $[C_nH_{2n-1}]^+$ ions are more prominent in olefins with large numbers of allylic hydrogens. Although a full range of alkyl and alkenyl ions are formed, the intensity distribution normally peaks in the C_4–C_7 region. The MH⁺ intensity is a maximum for C_6 olefins (15 to 20% of total ionization) and drops off with increasing size to 2 to 3% of total additive ionization. This intensity, combined with similar intensities for the [M – H]⁺ ion, is sufficient to establish molecular masses. However, it should be noted that the CH_4 CI mass spectra give no reliable indication of the position of the double bond (as is evident from Figure 1) or the extent and location of chain branching.

Budzikiewicz and Busker[12] have reported the i-butane CI mass spectra of a number of n-octadecenes. The spectra are similar to the CH_4 CI mass spectra in showing MH⁺, [M – H]⁺, alkyl, and alkenyl ions. In addition, a prominent $[M + C_4H_9]^+$ adduct ion is observed. The Z and E isomers were distinguishable from the $[M + C_4H_9]^+/[M – H]^+$ ratio which was ~2 for the Z isomers and <1 for the E isomers.

The NO CI mass spectra of various olefins have been reported;[9,13] in addition the NO CI mass spectra of the C_6 to C_9 n-olefins have been reported by Daishima et al.,[14] while the spectra of some internal n-alkenes have been

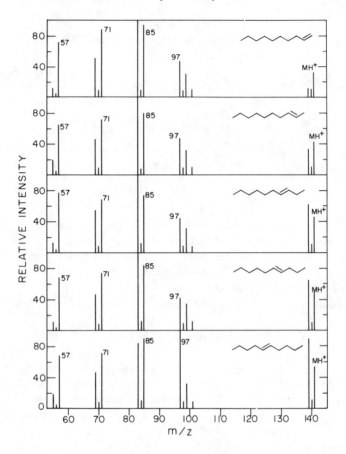

FIGURE 1. CH_4 CI mass spectra of isomeric n-decenes.

reported by Budzikiewicz and Busker.[12] Table 3 summarizes the NO CI mass spectra of the n-octenes reported by Daishima. For internal olefins, the spectra usually are dominated by $[M - H]^+$ ions, formed by hydride ion abstraction, and $[M + NO]^+$ adduct ions. If the double bond is highly substituted, the IE of the olefin will be lowered to the extent that charge exchange becomes exothermic and $M^{+\cdot}$ ions will be observed. For example, for 2,3-dimethyl-2-butene, $M^{+\cdot}$ carries about 40% of the total additive ionization.[13] For terminal olefins, only weak $[M - H]^+$ ion signals are observed although the $[M + NO]^+$ ion signal remains strong. In addition, a series of ions $C_nH_{2n}NO^+$ (n = 3 to 6) are observed at m/z values of 72, 86, 100, and 114, as shown in Table 3 for 1-octene. These ions are postulated[13] to arise by addition of NO^+ to the double bond followed by hydrogen migration and fragmentation as illustrated in Scheme 1 (R = H). These $C_nH_{2n}NO^+$ ions are formed to only a minor extent for internal olefins, although their abundance is enhanced at low source temperatures (see below).[12] Budzikiewicz and Busker[12] also noted that for a number of internal n-alkenes, the $[M + NO]^+/[M - H]^+$ ratio was greater (~2) for the Z isomers than

TABLE 3
NO Chemical Ionization Mass Spectra of n-Octenes

	[M–H]$^+$	M$^+$	C$_3$H$_6$NO$^+$	C$_4$H$_8$NO$^+$	C$_5$H$_{10}$NO$^+$	C$_6$H$_{12}$NO$^+$	[M+NO]$^+$
				Relative intensity			
1-Octene	11.8	0.6	9.6	100	70.3	11.3	31.3
cis-2-Octene	100	3.7	0.7	1.7	15.9	5.0	21.1
trans-2-Octene	100	2.2	—	1.0	6.9	2.3	9.2
cis-3-Octene	100	4.8	—	1.2	1.5	3.3	17.1
trans-3-Octene	100	2.7	—	0.6	0.5	1.0	6.0
cis-4-Octene	100	4.5	—	0.6	3.0	1.0	12.3
trans-4-Octene	100	2.8	—	—	0.9	—	4.1

for the E isomers (<1). This difference also is illustrated by the data for n-octenes (Table 3), although the $[M + NO]^+$ intensities are lower than they are for the larger alkenes.

In negative ion studies, the OH^- CI mass spectra of a few C_8–C_{10} alkenes have been reported by Smit and Field.[15] Although no $[M - H]^-$ ions were observed, low intensity $[M–3H]^-$ ions were present in all mass spectra. The major ions observed corresponded to $[M+43]^-$ and $[M+25]^-$; the former product was postulated to arise by reaction of the initial $[M - H]^-$ product with N_2O to form $[M - H + N_2O]^-$, which by elimination of H_2O gives the $[M+25]^-$ product. The adduct ions $[M+43]^+$ and $[M+25]^+$ also have been observed in the OH^- CI of terpene hydrocarbons[16] and cholestene,[17] although in both cases, $[M - H]^-$ ions were observed. In flowing afterglow studies, Bohme and Young[18] have observed $[M - H]^-$ ions as the product of reaction of OH^- with a number of small alkenes; their absence in the spectra reported by Smit and Field must be due to their reactivity with N_2O and with a lower proportion of N_2O in the reagent gas mix, $[M - H]^-$ ions might be observed in the CI mass spectra of alkenes.

A problem not resolved by electron ionization, Brønsted acid CI or Brønsted base CI is the location of the double bond in alkenes and other olefinic compounds. Budzikiewicz[19] has discussed in detail the approaches that have been taken to determine the double bond location directly by chemical ionization techniques. Some success has been achieved with simple olefinic compounds using the cycloaddition reaction of methyl vinyl ether discussed in Section II.D, Chapter 4; however, the method has seen only limited success in practical applications.[19–21] Budzikiewicz and Busker[12] have shown that the double bond location in Z-n-alkenes can be identified from the intensity distribution of the alkyl ions in the i-C_4H_{10} CI mass spectra. However, the method does not provide information on double bond location in E-n-alkenes, alkenes bearing functional groups, or dienes.

Although Hunt and Harvey[13] reported that N-containing fragment ions were observed only for terminal olefins in NO CI mass spectra, Budzikiewicz and co-workers[12,19] have shown that $C_nH_{2n}NO^+$ ions are observed at sufficiently low source temperatures for internal Z-n-alkenes with greater than 10 carbon atoms as well as for dienes and functionalized olefins. The ions are formed by the mechanism indicated in Scheme 1 (R = H or alkyl), and the intensities of these characteristic ions become negligible above ~140°C source temperature. These characteristic ions are not seen for E-alkenes nor for tri- and tetra-substituted alkenes. Einhorn and Malosse[22,23] have reported that olefinic compounds of the type $CH_3(CH_2)_xCH=CH(CH_2)_yR$ (R = CH_2OH, CH_2OAc, and CHO) give two acyl ions, $CH_3(CH_2)_xCO^+$ and $R(CH_2)_yCO^+$, under NO CI conditions; it has been shown that these products also are formed under suitable conditions for nonfunctionalized olefins.[24] Although the detailed mechanism of formation of these products is not clear,[25] there is some evidence that they originate by surface-catalyzed oxidative cleavage of the double bond to form the appropriate aldehydes which are subsequently ionized by hydride abstraction by NO^+. Since these products are observed only at low

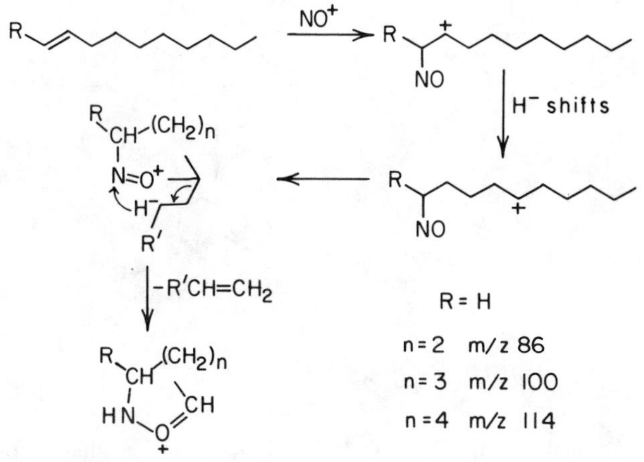

$$R\diagdown\!\!\!=\!\!\!\diagup \xrightarrow{NO^+} R\diagdown\overset{+}{\underset{NO}{C}}\diagup$$

H⁻ shifts

$$\underset{R'}{\overset{R\diagdown_{CH}\diagdown^{(CH_2)_n}}{\underset{\diagdown H\diagup}{N\!\equiv\!O^+}}} \quad\longleftarrow\quad R\diagdown\underset{NO}{\overset{+}{\diagup}}\diagdown$$

$\Big| -R'CH\!=\!CH_2$

R = H

$$\underset{HN\diagdown_{O}\diagup}{\overset{R\diagdown_{CH}\diagdown^{(CH_2)_n}}{\underset{+}{\diagdown CH}}}$$

n = 2   m/z 86
n = 3   m/z 100
n = 4   m/z 114

SCHEME 1

source temperature and with a small sample size, care must be taken in using this method of locating the double bond. Peake and Gross[26] have shown that the collision-induced dissociation of the Fe⁺ complex with alkenes leads to allylic cleavage of the alkene giving a Fe⁺-containing fragment ion characteristic of the double bond location. To date, only relatively small olefins have been studied in detail.

The CH_4 CI mass spectra of a variety of cycloalkanes have been determined by Field and Munson.[27] Like their alkene isomers, they show a series of alkyl and a series of alkenyl ions beginning, respectively, with the MH⁺ and [M – H]⁺ ions. Although, in many cases, the MH⁺ ion signal is low, the [M – H]⁺ ion signal usually comprises 25 to 75% of the total additive ionization.

Finally, the charge exchange mass spectra of isomeric C_5H_{10} compounds,[28] C_6H_{12} compounds,[29] and branched C_7H_{14} olefins[30] have been determined using reagent ions with recombination energies from 10.0 eV (CS_2^+) to 15.7 eV (Ar^+). Somewhat greater ability than EI to distinguish isomeric structures was noted.

IV. ALKYNES, ALKADIENES, AND CYCLOALKENES (C_nH_{2n-2})

The CH_4 CI mass spectra of a few alkynes, alkadienes, and cycloalkenes have been reported by Field,[11] while the i-C_4H_{10} CI mass spectra of some octadecynes have been reported by Busker and Budzikiewicz.[31] In the CH_4 CI mass spectra two series of ions were reported, the $[C_nH_{2n-1}]^+$ ions beginning with MH⁺ and the $[C_nH_{2n-3}]^+$ ions beginning with [M – H]⁺. For the alkynes and alkadienes, the former series predominated, while with the cycloalkenes, the [M – H]⁺ ion signal was particularly large. The i-C_4H_{10} CI mass spectra of the octadecynes showed the same two series of ions with the $[C_nH_{2n-1}]^+$ ions dominating. In addition, $[C_nH_{2n+1}]^+$ and $[C_nH_{2n-5}]^+$ ions were observed in low

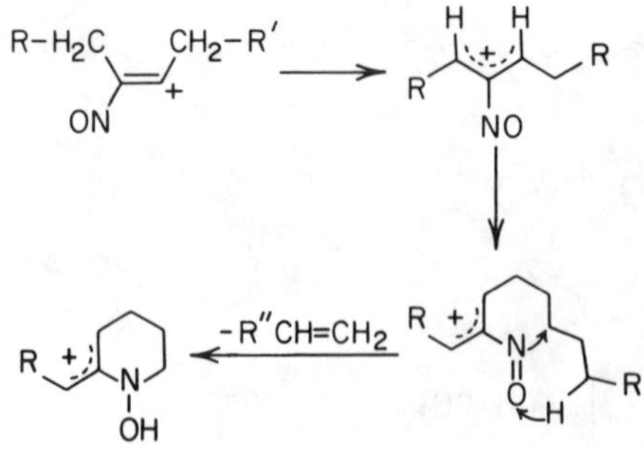

SCHEME 2

abundance as well as the $[M + C_4H_9]^+$ adduct ion. The distribution of the $[C_nH_{2n-1}]^+$ ions showed some correlation with the position of unsaturation.

Busker and Budzikiewicz[31] also examined the NO CI mass spectra of a number of dialkylacetylenes. Low intensity $[M + NO]^+$ (~10%) ion signals were observed, the major ions being the $[C_nH_{2n-3}]^+$ series accompanied by less abundant $[C_nH_{2n-1}]^+$ ions and relatively unimportant $[C_nH_{2n+1}]^+$ ion signals. In addition, N-containing ions characteristic of the triple bond were observed at low source temperatures but were observed to decrease rapidly in abundance with increasing source temperature. These ions consisted of two series $[C_nH_{2n}NO]^+$ and $[C_nH_{2n-2}NO]^+$ ions originating by cleavage of the $[M + NO]^+$ ion as indicated in Scheme 2. In each case, the cleavage is accompanied by migration of one additional H to the ionic portion. These ions allow location of the triple bond in dialkylacetylenes containing >10 carbon atoms including alkynes containing functional groups.[19] Busker and Budzikiewicz also examined the NO CI mass spectra of a number of 1-alkynes. For pent-1-yne and hex-1-yne the $[M + NO]^+$, $M^{+\cdot}$, and $[M - H]^+$ ions were of very low abundance, the major ions belonging to the $[C_nH_{2n-3}]^+$ series. Beginning with hept-1-yne ions of the series $[C_nH_{2n-2}NO]^+$ were observed.

The charge exchange mass spectra of the C_7 to C_9 n-alkynes have been determined using $CS_2^{+\cdot}$ and $N_2O^{+\cdot}$ as reactant ions;[32] isomer distinction was no more readily achieved from the charge exchange mass spectra than it was from the electron ionization mass spectra.

Smit and Field[15] reported the OH^- CI mass spectra of 1,7-octadiene, 1,4-octadiene, and 1-octyne. $[M - H]^-$ ions were observed in all cases along with $[M-3H]^-$ ions; the $[M - H]^-$ ion of 1,7-octadiene was particularly reactive with N_2O giving $[M - H + N_2O]^-$ and $[M - H + N_2O-H_2O]^-$.

V. AROMATIC HYDROCARBONS

In their early studies, Munson and Field[33] determined the CH_4 CI mass spectra of 21 alkylbenzenes and 2 alkylnaphthalenes. Several subsequent limited studies of the proton transfer CI of alkylbenzenes have been reported,[34-37] including a study of the C_8 to C_{10} alkylbenzenes using a variety of protonating agents

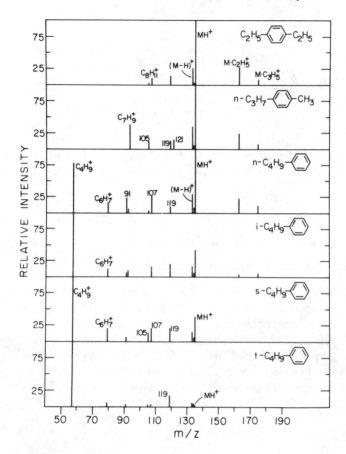

FIGURE 2. CH_4 CI mass spectra of C_{10} alkylbenzenes. Data from Herman and Harrison.[38]

in which the protonation exothermicity was systematically varied.[38] The gas phase ion chemistry of protonated alkylbenzenes recently has been reviewed with an emphasis on the unimolecular fragmentation reactions occurring on the metastable ion time scale.[39]

Aromatic molecules with no alkyl substituents show only MH^+ ions and the cluster ions $[M + C_2H_5]^+$ and $[M + C_3H_5]^+$. With the introduction of methyl groups, H^- abstraction also becomes possible and increases in importance as the number of methyl substituents increases; thus, pentamethylbenzene shows an $[M - H]^+$ ion signal which is 24% of the total additive ionization while toluene shows an $[M - H]^+$ ion signal which is only 3% of the total additive ionization.[33] With increasing size of the alkyl substituent(s), other reaction modes become possible. To illustrate, Figure 2 shows the CH_4 CI mass spectra of six C_{10} alkylbenzenes. A minor reaction channel is nominal alkide ion abstraction illustrated in Reaction 5 for s-butylbenzene; this reaction mode leads primarily to benzylic type ions at m/z values of 91, 105, 119, etc.

$$CH_5^+ + \text{(benzene ring)}-\overset{\overset{C}{|}}{C}-C-C \longrightarrow \begin{cases} \text{(benzene ring)}-\overset{\overset{C}{|}}{C}^+ + C_2H_6 + CH_4 \\ \\ \text{(benzene ring)}-\overset{+}{C}-C-C + CH_4 + CH_4 \end{cases} \tag{5}$$

The $C_2H_5^+$ ion reacts, in part, to form adduct ions, some of which fragment by olefin elimination from the alkyl substituent, Reaction 6. This displacement

$$C_2H_5^+ + \text{(benzene ring)}-\overset{\overset{C}{|}}{C}-C-C \longrightarrow \overset{C_2H_5^+}{\text{(benzene ring)}}-\overset{\overset{C}{|}}{C}-C-C \longrightarrow \overset{H^+}{\text{(benzene ring)}}-C_2H_5 + C_4H_8 \tag{6}$$

reaction leads to protonated ethylbenzenes; examples are the m/z 107 product in the mass spectra of the butylbenzenes and the m/z 121 product in the mass spectrum of n-propyltoluene (Figure 2).

However, the major reaction channel for both CH_5^+ and $C_2H_5^+$ continues to be proton transfer leading to MH⁺. The MH⁺ ion may fragment either by olefin elimination (Reaction 7) or by elimination of a neutral aromatic molecule with formation of the appropriate alkyl ion (Reaction 8). Reaction 7 is responsible

$$\overset{H^+}{\text{(benzene ring)}}-\overset{\overset{C}{|}}{C}-C-C \longrightarrow \text{(benzene ring)}---\overset{\overset{+\overset{C}{|}}{}}{C}-C-C \begin{cases} \longrightarrow \text{(ring)} + C_4H_8 \tag{7} \\ \\ \longrightarrow C_4H_9^+ + \text{(benzene ring)} \tag{8} \end{cases}$$

for the ions $C_6H_7^+$ (butylbenzenes), $C_7H_9^+$ (n-propyltoluene), and $C_8H_{11}^+$ (diethylbenzene) in the spectra of Figure 2. From isotopic labelling studies, Reaction 7 has been shown[35,37,40] to involve non-random transfer of hydrogen from several positions of the alkyl group rather than specific migration from a single position. There is substantial evidence[38–40] that Reactions 7 and 8 proceed, as shown, through the intermediacy of an ion/neutral complex consisting of an alkyl ion and a neutral benzene. Within this complex, hydrogen exchange may occur between the alkyl ion and the benzene involving the formation of a second ion/neutral complex consisting of a protonated benzene and an olefin.[40] In addition, within the complex, the alkyl ion may undergo hydride shifts to form a more stable alkyl ion; such shifts are necessary to rationalize the predominant formation of $C_4H_9^+$ rather than $C_6H_7^+$ in the spectra of n-butylbenzene and i-butylbenzene.[38] As the protonation exothermicity increases, the fragmentation pattern remains essentially the same and the MH$^+$ intensity decreases.[38] The $C_4H_9^+$ ion (i-C_4H_{10} CI) is reported to react only very slowly with aromatic hydrocarbons;[41] the proton transfer step is endothermic.

While proton transfer chemical ionization normally can distinguish structurally isomeric alkylbenzenes, it cannot distinguish positionally isomeric alkylbenzenes. Hawthorne and Miller[42] have shown that the aromatic hydrogens of alkylbenzenes are, in part, replaced by deuterium when CH_3OD is used as a reagent gas. The extent of exchange can be used to identify the number of substituents on the aromatic ring and, for disubstituted compounds, the *meta* isomer can be distinguished from the *ortho* and *para* isomers. Alkyl substitution on the aromatic ring of indans and tetralins can be distinguished from alkyl substitution on the non-aromatic ring in the same way.[43] The subject of isotopic exchange reactions will be discussed in more detail in Chapter 6.

The NO CI mass spectra of a few substituted benzenes, including alkylbenzenes, have been recorded.[9,44,45] The [M + NO]$^+$ adduct ion is the dominant product for alkylbenzenes having ionization energies greater than ~8.7 eV, while M$^+$ is dominant for those with ionization energies which are lower than ~8.7 eV. The NO$^+$ ion reacts only to a minor extent with alkylbenzenes by H$^-$ abstraction.

The OH$^-$ CI mass spectra of a number of alkylbenzenes have been recorded by Smit and Field.[15] Benzene does not react with OH$^-$ since it has no acidic hydrogens. However, the benzylic hydrogens of alkylbenzenes are sufficiently acidic to be abstracted by OH$^-$, and [M − H]$^-$ is the primary reaction product. The [M − H]$^-$ ions react, in part, to add N_2O (from the reagent mixture) to form [M+43]$^-$, some of which fragment by loss of H_2O to give [M+25]$^-$.

The reactions of O$^-$ with a variety of aromatic hydrocarbons have been studied by medium pressure techniques.[46-49] The products observed indicate the following sequence of reactions where R is a methyl group and X is a halogen or hydrogen atom. Benzene undergoes

$$O^{-} + M \rightarrow [M - H]^{-} + OH^{\cdot} \tag{9}$$

$$\rightarrow [M - 2H]^{-\cdot} + H_2O \tag{10}$$

$$\rightarrow [M - H + O]^{-} + H^{\cdot} \tag{11}$$

$$\rightarrow [M - R + O]^{-} + R^{\cdot} \tag{12}$$

$$\rightarrow [M - H + O - HX]^{-} + H^{\cdot} + HX \tag{13}$$

TABLE 4
O⁻ Chemical Ionization Mass Spectra of C_8H_{10} Alkylbenzenes

Isomer	Relative intensity			
	[M – H + O]⁻	[M–CH₃+O]⁻	[M – H]⁻	[M – 2H]⁻
o-Xylene	42	17	100	56
m-Xylene	9	4	23	100
p-Xylene	74	19	100	86
Ethylbenzene	100	—	96	40

only Reactions 10 and 11 giving peaks at m/z 76 and m/z 93, respectively, the latter accounting for about ²/₃ of the additive ionization. With naphthalene, Reaction 11 was by far the most important process while with anthracene, both H displacement and simple charge exchange to give M⁻ were observed. The O⁻ CI mass spectra of alkylaromatic compounds are illustrated by the spectra of the isomeric C_8H_{10} alkylbenzenes given in Table 4.[49] All the products of Reactions 9 to 12 are observed, the relative intensities depending on molecular structure. An additional product which results from reaction of O⁻ is OH⁻. This species may react further with the alkylbenzene to produce [M – H]⁻; consequently, the spectra will depend on sample size.[49] The spectra reproduced in Table 3 were recorded with similar sample sizes.

Polycyclic aromatic hydrocarbons (PAHs) have been extensively studied since they are potential environmental pollutants and many are carcinogenic. The CH_4 CI mass spectra show M⁺·, MH⁺, [M + C_2H_5]⁺, and [M + C_3H_5]⁺; when H_2 is used as reagent only the M⁺· and MH⁺ ions are observed.[50] The IE of many PAHs are quite low so that charge transfer from the reagent ions may occur; in addition, it has been pointed out[51] that in many cases residual electron ionization cannot be neglected. CH_4 CI gave a response factor (signal per nanogram of sample) which was less than that for EI while H_2 CI gave a response factor similar to EI.

Simonsick and Hites[52] have used a reagent mixture of 15% CH_4 in argon and observed both proton transfer and charge exchange ionization. They reported that the ratio [M + H]⁺/M⁺· increased with an increase in the IE of the PAH, and they used the measured ratios to differentiate isomeric PAHs. This approach requires careful control of experimental conditions; in the absence of such control, Brotherton and Gulick[50] did not observe such a correlation

A number of workers have examined the electron capture ionization of PAHs.[50,53-55] As discussed in Section III.A, Chapter 4, widely differing sensitivities or response factors are observed. In general, M⁻· or [M – H]⁻ or both were observed, although in many cases, there are substantial variations in the spectra reported from the different laboratories, and artifact peaks are reasonably common. Oehme[54] also reported the negative ion CI mass spectra of a selection of PAHs using CH_4/N_2O as reagent gas (OH⁻ as major reactant).

The ions observed were [M − H]⁻, M⁻·, [M + OH]⁻, [M − H + NO]⁻ and [M − H + N₂O]⁻. The formation of M⁻ is unexpected since none of the PAHs studied have EAs near that of OH⁻; it is possible that there was a substantial concentration of electrons in the reagent plasma. The response factors were reported to be higher than for EI. Brotherton and Gulick[50] have reported the CI mass spectra obtained using N_2/N_2O as a reagent mixture where O⁻ should be the dominant reagent species. [M–2H]⁻·, [M − H]⁻, M⁻·, and [M − H + O]⁻ were the major ions observed; again the formation of M⁻ is surprising. The response factors were generally lower than those obtained using EI although some of the compounds which gave a high response by electron capture also gave a very high response by O⁻ CI, suggesting the possibility that a considerable population of thermal energy electrons might be present in the plasma.

VI. ALCOHOLS

The i-C_4H_{10} CI mass spectra of 23 saturated, monohydroxylic alcohols have been reported by Field,[56] while partial CH_4 CI mass spectra of a variety of saturated, olefinic and cyclic alcohols have been reported by Sarris et al.[57] Winkler and co-workers have reported detailed studies of the isobutane[58] and ammonia[59] CI mass spectra of a variety of alcohols, while Munson et al.[60] have reported the isobutane CI mass spectra of a number of unsaturated alcohols. In addition, the H_2 and CH_4 CI mass spectra of a series of substituted benzyl alcohols have been reported[61] as well as a detailed study of isomeric C_5 and C_6 alkanols using H_3^+, N_2H^+, CO_2H^+, N_2OH^+ and HCO^+ as protonating agents.[62] The isobutane CI mass spectra of a number of terpene alcohols also have been determined.[16,63]

The general features of the Brønsted acid chemical ionization mass spectra of alkanols are illustrated in Figure 3 using 1-decanol as an example. For alkanols (ROH) higher than C_4, the ROH_2^+ ions are not stable and form R⁺ ions by loss of H_2O. Proton transfer from $C_4H_9^+$ to alkanols in i-C_4H_{10} CI is endothermic and it is not entirely clear how the R⁺ ions (m/z 141 of Figure 3) originate. The other primary reaction channel is hydride abstraction to form [M − H]⁺ (m/z 157 of Figure 3); this product is not observed in the CI mass spectra of tertiary alkanols since the hydride abstracted originates from the carbon to which the oxygen is attached. Both the [M − H]⁺ and R⁺ ions undergo further fragmentation, the extent of which depends on the exothermicity of the initial ionization reaction;[62] this is illustrated by the lower intensity of low mass ions in the i-C_4H_{10} CI mass spectrum of Figure 3. The major fragmentation reactions of the R⁺ alkyl ions involve elimination of olefins while the [M − H]⁺ ion eliminates H_2O to form an alkenyl ion, [R–H₂]⁺, which may fragment further by olefin elimination. Thus a series of alkyl and alkenyl ions are observed at lower masses. The sensitivity for ammonia CI of alkanols is reported to be low[57] and the spectra observed are in many cases complex,[59]

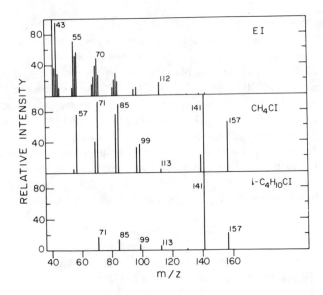

FIGURE 3.   EI and CI mass spectra of n-decanol. CH_4 CI data from
Munson and Field.[64] i-C_4H_{10} CI data from Field.[56]

involving in some cases formation of RNH_3^+ by displacement of the hydroxyl
group and formation of cluster ions.

The MH^+ of simple cyclic alcohols, allylic alcohols, and benzylic alcohols
are unstable with respect to fragmentation by loss of H_2O. Indeed, MH^+ ions
are only observed when there is another functional group present in the
molecule to stabilize the MH^+ ion by interaction with the protonated hydroxyl
group. Thus, in contrast to the results of Figure 3, the MH^+ ion constitutes
~14% of the total additive ionization in the CH_4 CI mass spectrum of 1,10-
decanediol and ~67% in the i-C_4H_{10} CI mass spectrum.[65] As a result of such
stabilization, MH^+ ions are more intense in the i-C_4H_{10} CI mass spectra of the
cis 1,3- and 1,4-cyclohexanediols than in the CI mass spectra of the *trans*
isomers.[66,67] Double bonds also can stabilize the protonated species. Thus,
while MH^+ is of very low abundance in the i-C_4H_{10} CI mass spectrum of hex-
2-en-1-ol, it is the base peak in the spectra of hex-3-en-1-ol, hex-4-en-1-ol, and
hex-5-en-1-ol.[60] Such double bond stabilization leads to abundant MH^+ ions in
the i-C_4H_{10} CI mass spectra of a number of terpene alcohols[16,63,68] and to some
pronounced stereochemical effects;[60,69] for example, the MH^+ ion is much
more abundant in the i-C_4H_{10} CI mass spectrum of *syn*-7-hydroxynorborn-2-
ene than it is in the spectrum of the *anti* isomer. Stereochemical effects will be
discussed in more detail in Chapter 6.

The NO CI mass spectra of a number of alkanols have been determined.[14,45,70] Tertiary alcohols show only an $[M - OH]^+$ ion signal. Secondary
alcohols show $[M - H]^+$, $[M - OH]^+$ and, $[M - 2H + NO]^+$ ion signals, while
primary alcohols show $[M - H]^+$, $[M - OH]^+$, $[M - 3H]^+$, and $[M-2H + NO]^+$

ion signals. The $[M - 2H + NO]^+$ product for secondary alcohols presumably arises from oxidation of the alcohol to a ketone followed by addition of NO^+. A similar oxidation of a primary alcohol leads to an aldehyde with which NO^+ reacts both by hydride ion abstraction and NO^+ addition. Although these reactions serve to distinguish primary and secondary alcohols, the extent of reaction is very dependent on experimental conditions; for example, the oxidation products are not observed when NO/H_2 mixtures are used to produce the NO^+.[9]

The reactions of OH^- with a number of alcohols have been investigated using both medium pressure CI techniques[15] and ICR techniques.[71] The dominant reaction product was found to be $[M - H]^-$ with, in a number of cases, additional formation of $[M - 3H]^-$ by elimination of H_2 from the $[M - H]^-$ ion. The OH^- CI mass spectra of terpene alcohols show $[M - H]^-$ and , in some cases, $[M - 3H]^-$ ions; in a number of unsaturated terpene alcohols, elimination of H_2O from the $[M - H]^-$ ion also was observed.[16] Much more extensive fragmentation was observed in the O^- CI of the few alcohols examined by Houriet et al.,[71] Table 5 records the spectra reported for a selection of alcohols. At first glance the higher extent of fragmentation is surprising since O^- has a similar PA as OH^- (Table 8, Chapter 4). The formation of the additional fragment ions was rationalized by the mechanism shown in Scheme 3, where R' may be a hydrogen or an alkyl group. As a consequence of this reaction pattern, O^- CI mass spectra provide information on the branching at the α-carbon.

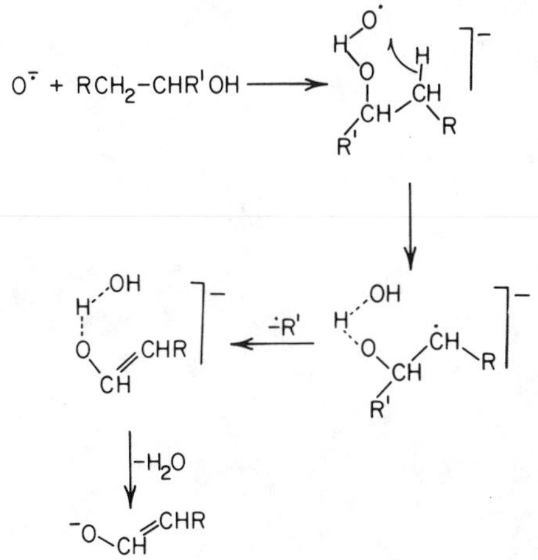

SCHEME 3

Although ECCI of alcohols is not feasible, electrophoric derivatives which undergo electron capture and provide molecular mass information have been studied.[72]

TABLE 5
O⁻ Chemical Ionization of Alkanols

M			Relative intensity		
	$[M-H]^-$	$[M-3H]^-$	$[M-(H_2+CH_3)]^-$	$[M-(H_2+C_2H_5)]^-$	$[M-(H_2+C_3H_7)]^-$
$CH_3CH(OH)CH_3$	100	21	34	—	—
$CH_3(CH_2)_3OH$	100	37	—	—	—
$(CH_3)_2CHCH_2OH$	100	43	—	—	—
$C_2H_5CH(OH)CH_3$	100	27	23	58	—
$(CH_3)_3COH$	59	—	100	—	—
$i\text{-}C_3H_7CH(OH)CH_3$	100	13	22	—	48
$C_3H_7C(OH)(CH_3)C_2H_5$	77	—	79	100	87

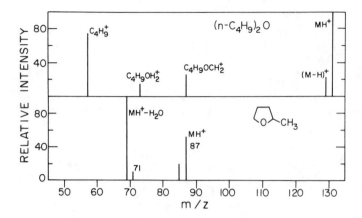

FIGURE 4. CH$_4$ CI mass spectra of di-n-butyl ether and 2-methyltetrahydrofuran.

VII. ETHERS

Alkyl ethers have not been studied extensively by chemical ionization techniques. The CH$_4$ CI mass spectrum of di-n-butyl ether is shown in Figure 4 as a representative ether. In contrast to the analogous alcohol, where MH$^+$ is absent, the MH$^+$ ion comprises the base peak in the spectrum of the ether. A moderate intensity [M – H]$^+$ ion signal also is observed in the molecular mass region. The major fragment ion is the alkyl ion formed by alcohol elimination from MH$^+$; olefin elimination from the [M – H]$^+$ ion to form C$_3$H$_7$CHOH$^+$ (m/z 73) also is observed, as is nominal alkide ion abstraction to form C$_4$H$_9$OCH$_2^+$.

The i-C$_4$H$_{10}$ CI mass spectra of a number of cyclic ethers have been reported;[73] abundant MH$^+$ ions are observed with the major fragmentation channel involving loss of H$_2$O from MH$^+$. This fragmentation mode also is the major decomposition channel for the MH$^+$ ions of 2-methyloxetan and tetrahydrofuran[74] as well as tetrahydropyran, 2-methyltetrahydrofuran and 3-methyltetrahydrofuran.[75] The CH$_4$ CI mass spectrum of 2-methyltetrahydrofuran is shown in Figure 4. Elimination of H$_2$O also is the most common fragmentation reaction of protonated aldehydes, and it is probable that the protonated cyclic ether rearranges to a protonated aldehyde prior to fragmentation. The mechanism of elimination of H$_2$O clearly is complex since neither for the cyclic ethers[74,75] nor for the aldehydes[75,76] is the added proton always lost with the H$_2$O eliminated. The MH$^+$ ion of phenyl alkyl ethers fragment by olefin elimination to form the protonated phenol if the alkyl group is larger than methyl.[77]

The NO CI mass spectra of the few open chain and cyclic ethers that have been studied[9,45] showed [M – H]$^+$ as the sole product ion.

The reaction of OH$^-$ with dialkyl ethers R$_2$O produces RO$^-$ with some loss of H$_2$ from the alkoxide ion.[15]

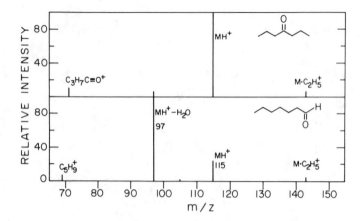

FIGURE 5.   CH₄ CI mass spectra of 4-heptanone and n-heptaldehyde.

VIII. ALDEHYDES AND KETONES

The proton transfer chemical ionization mass spectra of a selection of C₃[78], C₄[74], and C₅[75] carbonyl compounds have been determined using protonating agents ranging from H₃⁺ to HCO⁺; in addition, the H₂ and CH₄ CI mass spectra of eight C₆ carbonyl compounds have been reported.[79] The CH₄ CI mass spectrum of n-heptanal also has been reported.[76] The CH₄ CI mass spectra of 4-heptanone and n-heptanal are shown in Figure 5. For the ketone, MH⁺ is the base peak with only minor fragmentation; this seems to be generally true for ketones in CH₄ CI. By contrast, protonated aldehydes show much more facile fragmentation by elimination of H₂O; as mentioned above, the detailed mechanism is complex and the added proton is not always lost with the H₂O eliminated.[75,76,79] The fragmentation of the protonated ketones can be induced by more exothermic protonation;[75,79] the fragmentation reactions observed depend strongly on the precursor structure and there is significant rearrangement of the molecular structure prior to fragmentation.

The NO CI mass spectra of a number of aldehydes and ketones have been reported.[9,45,80] The spectra of the aldehydes normally show [M – H]⁺ ions and either [M + NO]⁺ or M⁺˙ ions depending on the ionization energy of the aldehyde. Ketones do not show [M – H]⁺ ions but only [M + NO]⁺ or M⁺˙ ions, again depending on the IE of the ketone. In both cases, charge exchange is not important unless the IE of the substrate molecule is below ~8.5 to 8.8 eV.

The OH⁻ CI mass spectra of a few ketones have been reported.[15] The major product ion is [M – H]⁻ with a minor yield of [M–3H]⁻.

The reactions of O⁻ with a variety of carbonyl compounds have been investigated by ICR techniques;[81] more recently, the O⁻ CI mass spectra of twenty-eight C₄ to C₇ carbonyl compounds have been reported.[82] Table 6 records the O⁻ CI mass spectra of six C₆ ketones. The primary reactions of O⁻ with ketones are as follows:

$$O^{-} + RCOR' \rightarrow [M-H]^{-} + OH^{-} \tag{14}$$

$$\rightarrow [M-H]^{-} + OH^{\cdot} \tag{15}$$

$$\rightarrow [M-2H]^{-\cdot} + H_2O \tag{16}$$

$$\rightarrow RCOO^{-} + R'^{\cdot} \tag{17}$$

$$\rightarrow R'COO^{-} + R^{\cdot} \tag{18}$$

It has been shown[82,83] that the two hydrogens abstracted in the H_2^+ abstraction, Reaction 16; may originate from the same α-carbon atom or one may originate from each α-carbon atom. The $[M-2H]^{-}$ ions so formed also undergo further fragmentation as indicated in Reactions 19 to 21 for the $[M-2H]^{-}$ ions derived from 2-hexanone.

$$^{-\cdot}CHC(=O)CH_2CH_2CH_2CH_3 \rightarrow HCCO^{-} + C_4H_9 \tag{19}$$

$$CH_3C(=O)C^{-\cdot}CH_2CH_2CH_3 \rightarrow C_3H_7CCO^{-} + CH_3^{\cdot} \tag{20}$$

$$^{-}CH_2C(=O)CH^{\cdot}CH_2CH_2CH_3 \rightarrow {}^{-}CH_2C(=O)CH=CH_2 + C_2H_5^{\cdot} \tag{21}$$

The O^{-} CI mass spectra provide considerable information as to the structure of the ketone and some isomer differentiation as is evident from the spectra summarized in Table 6. For cases where the spectra are similar, as for example, 3-hexanone and 2-methyl-3-pentanone, it has been shown that the collisional charge inversion mass spectra of the $[M-H]^{-}$ ions allows isomer differentiation.[82] The O^{-} CI mass spectra of aldehydes, RCHO, show $[M-H]^{-}$ and the two carboxylate ions HCO_2^{-} and RCO_2^{-}, the latter being formed preferentially.

IX. CARBOXYLIC ACIDS AND ESTERS

The CH_4 CI mass spectra of carboxylic acids normally show appreciable MH^+ ion signals with the dominant fragmentation reaction being loss of H_2O from MH^+.[61,84,85] Benzoic acids also show loss of CO_2 from MH^+ resulting in a protonated benzene ion.[61] The elimination of H_2O is enhanced substantially by interaction with a second carboxyl group. Thus, dodecanoic acid shows an $[MH-H_2O]^+/MH^+$ ratio of 0.7 while dodecan-1,10-dioic acid shows a ratio of 50.[84] Similarly, fumaric acid, where the two carboxyl groups cannot interact, shows a $[MH-H_2O]^+/MH^+$ ratio of 0.2 compared to a ratio of 10 for maleic acid.[85] In the same vein, H_2O loss is much more pronounced from the MH^+ ion of phthalic acid than from the MH^+ ion of isophthalic acid.[84] In protonated benzoic acids loss of H_2O is more pronounced in those acids bearing an *ortho* substituent with a lone pair of electrons, e.g., OH, NH_2, F, Cl.[61]

The thermochemically favored site of protonation of carboxylic acids is

TABLE 6
O⁻ Chemical Ionization Mass Spectra of C₆ Ketones

m/z	Ion	Relative intensity					
		2-Hexanone	3-Methyl-2-pentanone	4-Methyl-2-pentanone	3,3-Methyl-2-butanone	3-Hexanone	2-Methyl-3-pentanone
101	$C_4H_9CO_2^-$	15	15	11	14	100	100
99	$[M–H]^-$	100	100	100	100		
98	$[M–2H]^-$	59		65			
87	$C_3H_7CO_2^-$					23	37
83	$[M–2H–CH_3]^-$	24		55			
73	$C_2H_5CO_2^-$					22	31
69	$[M–2H–C_2H_5]^-$	18					
59	$CH_3CO_2^-$	35	15	9	12		
41	$HCCO^-$	16	38	11	48		

the carbonyl carbon.[86] Loss of H_2O from this species (Reaction 22) requires a 1,3-H shift which is symmetry-forbidden and has a large energy barrier.[87]

$$R - \overset{\overset{\displaystyle OH}{|.\,+}}{\underset{}{C}} \overset{}{\underset{}{\text{\;} OH}} \longrightarrow R - \overset{\overset{\displaystyle O}{||}}{\underset{}{C}} - O\overset{+}{H_2} \longrightarrow RC{\equiv}O^+ + H_2O \qquad (22)$$

A second interacting carboxyl group obviates the necessity for such a 1,3-H shift and provides a lower energy route to fragmentation (Reaction 23).

$$(23)$$

In addition, the product likely is stabilized further by formation of the protonated anhydride. Similarly, in protonated benzoic acids, an *ortho* substituent capable of accepting a proton acts as an intramolecular catalyst for the hydrogen shift necessary for H_2O elimination (Reaction 24).

$$+ H_2O \quad (24)$$

The Brønsted acid CI mass spectra of a variety of methyl esters have been reported.[84,85,88,89] With CH_4 as reagent gas, MH^+ ions are observed as well as the adduct ions $[M + C_2H_5]^+$ and $[M + C_3H_5]^+$. The major fragmentation channel of MH^+ is elimination of CH_3OH and this fragmentation is similarly catalyzed by interaction with a second carboxyl group.[84,85] With increasing length of the alkyl chain, reactions characteristic of an alkane, i.e., H^- abstraction, are observed to increase in importance; in addition, loss of CH_3OH from $[M - H]^+$ is observed. Figure 6 shows the dependence of ion abundances on chain length in the H_2 CI of methyl n-alkanoates; the CH_4 CI system shows a similar behavior except that the abundances of the fragment ions are lower and the $[M + C_2H_5]^+$ and $[M + C_3H_5]^+$ cluster ions are observed.[89] .

The Brønsted acid CI mass spectra of a number of alkyl formates and acetates,[90,91] alkyl propionates,[92] n-alkyl hexanoates,[93] n-pentyl esters,[93] and long-chain wax esters[94] have been reported. With CH_4 as the reagent gas, significant intensities for the MH^+ ion are observed while with i-C_4H_{10} as

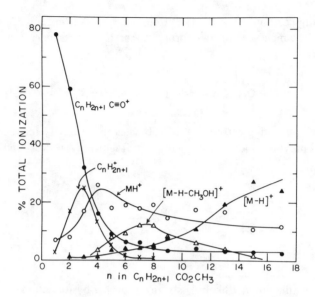

FIGURE 6.   Dependence of ion abundances on chain length in the H_2 CI mass spectra of methyl n-alkanoates. (From Tsang, C.W. and Harrison, A.G., *J. Chem. Soc. Perkin Trans. 2*, 1718, 1975. With permission.)

reagent, the MH$^+$ ions dominate the spectra. Alcohol loss from MH$^+$ is of only minor importance since two new lower energy reaction channels, elimination of an alkene (Reaction 25) and elimination of a carboxylic acid (Reaction 26), become possible. If the protonation step is sufficiently exothermic, fragmentation of the alkyl ion R$^+$ may occur.[91]

$$R'-\overset{\underset{\displaystyle k+}{OH}}{C}=OR \longrightarrow R'-\overset{\displaystyle OH}{\underset{}{C}}=O\cdots\overset{+}{R}$$

$$R'-\overset{\underset{\displaystyle k+}{OH}}{C}=OH + [R-H] \qquad (25)$$

$$R'-\overset{\displaystyle O}{\underset{\|}{C}}-OH + R^+ \qquad (26)$$

The CH$_4$ CI mass spectra of isopentyl and n-pentyl propionates (Figure 7) illustrate the major features of the Brønsted acid CI mass spectra of alkyl esters. The competition between Reactions 25 and 26 depends on the identity of the ester group and the alcohol group. For acetates, at least, straight chain R groups lead to dominant formation of the protonated acid while branched chain R groups lead to dominant formation of R$^+$ in fragmentation of MH$^+$.[91] Presumably, in the branched isomers, hydrogen shifts in the alkyl group permit formation of the more stable tertiary carbenium ion while with linear isomers, only the less stable secondary carbenium ions can be formed. As can be seen in Figure 7, the C$_2$H$_5^+$ and C$_3$H$_5^+$ ions react, in part, by addition and elimination, Reactions 27 and 28; these reactions are observed for propyl and higher esters.

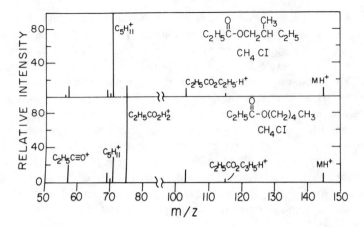

FIGURE 7.   CH$_4$ CI mass spectra of isopentyl and n-pentyl propionates.

$$C_2H_5^+ + R'CO_2R \rightarrow R'CO_2R^{\cdot}C_2H_5^+ \rightarrow R'CO_2C_2H_5H^+ + [R-H] \quad (27)$$

$$C_3H_5^+ + R'CO_2R \rightarrow R'CO_2R^{\cdot}C_3H_5^+ \rightarrow R'CO_2C_3H_5H^+ + [R-H] \quad (28)$$

The CH$_4$ and i-C$_4$H$_{10}$ CI mass spectra of a series of alkyl carbamates (R' = NH$_2$) have been reported;[95] MH$^+$ ions are observed, and the major features of the mass spectra are rationalized by Reactions 25 to 28. By contrast, the Brønsted acid CI of a variety of carbamate pesticides in which the R group was aromatic showed ROH$_2^+$ as the major fragment ion.[96] In contrast to alkyl alkanoates and carbamates, terpene esters[16,63,97] and cholesteryl esters[98,99] do not, in general, show MH$^+$ ions in their CH$_4$ or i-C$_4$H$_{10}$ CI mass spectra. For the terpene esters, the sole ion observed is the R$^+$ ion resulting from elimination of a neutral carboxylic acid; the one exception appears to be for citronellyl esters where the double bond in the hydrocarbon chain apparently interacts with the protonated ester function to stabilize the MH$^+$ ion which is observed. For cholesteryl esters both Reactions 25 and 26 are observed. The NH$_3$ CI of terpene and cholesteryl esters results primarily in formation of the ammonium ion adduct.[97,99] The CIMS of steroids and other lipids has been reviewed.[100]

The CI mass spectra of a number of dicarboxylic acids and their methyl esters have been determined with the reagent ions H$_3$O$^+$, CH$_3$OH$_2^+$, and NH$_4^+$, in each case using a mixture of the appropriate neutral in methane as reagent gas,[101] while the chemical ionization mass spectra of a variety of dicarboxylic acids have been obtained using pure CH$_3$OH as reagent gas.[102] An interesting feature of the spectra obtained in the two studies was the observation of exchange of the reagent gas for water or methanol of the dicarboxylic compound. Thus, sebacic acid, HO$_2$C(CH$_2$)$_8$CO$_2$H, showed MH$^+$, [MH – H$_2$O + CH$_3$OH]$^+$, and [MH – 2H$_2$O + 2CH$_3$OH]$^+$, with the latter ion being the most intense ion in the spectrum. In the latter work,[102] it was shown that to esterify both carboxyl groups they must be able to interact with each other; thus, fumaric acid showed incorporation of only one methanol while maleic acid showed some incorporation of two methanols. The H_2O chemical ionization mass spectra of a variety of esters show variable hydrolysis of the ester.[103]

The NO CI of esters leads to formation of $[M + NO]^+$ with some formation of acyl ions resulting from nominal alkoxide ion abstraction by NO^+.[9]

The OH^- CI mass spectra of carboxylic acids show $[M - H]^-$, presumably the carboxylate anion, as the sole product ion.[15] The Cl^- ion is reported to react with a number of dicarboxylic acids by proton abstraction to give $[M - H]^-$.[104] By contrast, Br^- reacted with some dicarboxylic acids by proton abstraction but with others by formation of $[M + Br]^-$;[104] for example, phthalic acid reacted by proton abstraction while isophthalic acid and terephthalic acid reacted primarily by adduct formation. It appears that interaction with a second carboxyl group stabilizes the anion sufficiently to make proton abstraction by the weak base Br^- possible.

The OH^- CI mass spectra of a variety of esters have been reported.[15-17,105-107] In general, for the ester $R'CO_2R$, $[M - H]^-$ and $R'COO^-$ ions are observed; in addition, the $[M - H]^-$ ion may eliminate an alcohol molecule, ROH. The report[15] that t-pentyl acetate formed primarily the $C_5H_{11}O^-$ alkoxide ion has been shown[16] to be in error. Eugenyl valerate, the ester of an aromatic alcohol, does form[16] primarily RO^- and this may be a general observation for the esters of aromatic alcohols. Cholesteryl esters give a low yield of an ion of m/z 367 corresponding, nominally, to elimination of $R'CO_2H$ from the $[M - H]^-$ ion.[17] The negative ion chemical ionization mass spectra of a variety of cholesteryl esters have been reported[108] using NH_3 as reagent gas where the ionization may be by electron capture or by reaction of NH_2^-. No $[M - H]^-$ ions were observed, the major ions being $R'COO^-$ and $[R'COO - H_2O]^-$ with a minor yield of m/z 367. The reactions of F^- and NH_2^- with acetate esters[109] are similar to the reactions of OH^- with the exception that the reaction of F^- leads, in part, to formation of $FCOCH_2^-$ by elimination of an alcohol molecule from the tetrahedral intermediate formed by attack of F^- on the carbonyl carbon.

Negative ion chemical ionization may be useful in the identification of the fatty acid esters of glycerol and related compounds. It has been reported[110] that the OH^- CI mass spectrum of mixed fatty acid esters of chloropropandiol show low intensity $[M - H]^-$ ions and intense carboxylate ions of the fatty acids. Recent work in this laboratory[111] has shown that the same is true for triacylglycerols; for example, the OH^- CI mass spectrum of 1,2-distearoyl-3-palmitoyl-*rac*-glycerol shows $[M - H]^-$ (10%), the stearate carboxylate ion (55%), and the palmitate carboxylate ion (35%). Very recently, it has been reported[112] that Cl^- attachment is a sensitive method for detecting triacylglycerols and the trimethylsilyl or t-butyldimethylsilyl ethers of diacylglycerols; by contrast, Cl^- attachment to cholesteryl esters was very inefficient. Some limited success in the analysis of intact phospholipids has been achieved using positive ion CI with ammonia as the reagent gas.[113,114]

ECCI of acids and esters appears to be relatively unsuccessful.[115] However, nitrobenzyl and pentafluorobenzyl derivatives of fatty acids undergo electron capture to afford the carboxylate anion with little fragmentation.[116,117]

X. AMINES

In terms of the discussion of Section II.A., Chapter 4, loss of NH_3 from $RNH_2.H^+$ is not a facile process, and appreciable MH^+ ion intensities should be seen in proton transfer CI mass spectra of amines. In line with these predictions, the CH_4 CI mass spectrum of cyclohexylamine shows MH^+ (100%), $[M - H]^+$ (70%), and $[MH - NH_3]^+$ (59%).[118] Similarly, in the CH_4 CI mass spectra, the $[MH - NH_3]^+/MH^+$ ion intensity ratios were reported to be 0.04 for n-butylamine, 0.07 for s-butylamine, and 2.3 for t-butylamine;[119] the high stability of the t-butyl carbenium ion overcomes the poor leaving tendency of NH_3. No other details of the CI mass spectra were given. When a second amine function is suitably located, as in α,ω-diaminobutane ($[MH - NH_3]^+/MH^+ = 2.3$), anchimeric assistance promotes the loss of NH_3.[119] In α,ω-aminoalcohols, H_2O is much more readily lost than NH_3 is following Brønsted acid CI.[119,120] Similar results have been obtained in the i-C_4H_{10} CI mass spectra of a variety of β-aminoalcohols.[121–123]

The CI mass spectra of a selection of C_3 to C_5 alkylamines have been determined using H_3^+ and HCO^+ as protonating agents.[124] The spectra of a selection of C_4 amines obtained with HCO^+ as protonating agent are presented in Table 7; CO has a PA similar to that of CH_4. Despite the poor leaving ability of NH_3, only low intensity MH^+ ion signals are observed for the butylamines, although much more intense MH^+ ion signals are observed for the secondary and tertiary amines. The two major fragmentation reactions of MH^+ are

$$RNH_3^+ \rightarrow R^+ + NH_3 \qquad (29)$$

$$\rightarrow NH_4^+ + [R - H] \qquad (30)$$

Protonated diethylamine and ethyldimethylamine show loss of C_2H_4 to form $C_2H_5NH_3^+$ and $(CH_3)_2NH_2^+$, respectively. There also are very abundant ion signals arising, nominally, by alkide and hydride ion abstraction from the amine. These reactions lead to iminium ions, for example $C_2H_5NHCH_2^+$ for diethylamine. Although these iminium ions could arise by alkane elimination from MH^+, theoretical calculations show[124] that alkane elimination has a high energy barrier, even though the reaction is the thermochemically most favorable. Alkane elimination reactions are observed in the high energy (8 keV) collision-induced dissociation mass spectra of protonated alkylamines, but are of only minor importance in the low energy (50 eV) collision-induced dissociation mass spectra.[125] Formation of iminium ions also was observed as a major process in the CH_4 CI mass spectra of some underivatized tricyclic amines.[126] Whitney et al.,[127] have reported the CH_4 and i-C_4H_{10} CI mass

TABLE 7
CO/H$_2$ Chemical Ionization Mass Spectra of C$_4$ Amines

m/z	Ion	Relative intensity					
		n-C$_4$H$_9$NH$_2$	i-C$_4$H$_9$NH$_2$	s-C$_4$H$_9$NH$_2$	t-C$_4$H$_9$NH$_2$	(C$_2$H$_5$)$_2$NH	(CH$_3$)$_2$NC$_2$H$_5$
74	MH$^+$	15	9	3	1	100	61
73	M$^+$	18	15	2	1	71	100
72	(M–H)$^+$	27	28	19	4	79	92
59	MH$^+$–CH$_3$	—	1	1	2	2	2
58	MH$^+$–CH$_4$	11	23	26	75	99	69
57	C$_4$H$_9^+$	95	100	65	100	4	3
56	C$_4$H$_8^+$	1	3	—	1	4	2
55	C$_4$H$_7^+$	4	7	6	4	—	—
46	MH$^+$–C$_2$H$_4$	(18)^a	(10)^a	(16)^a	(3)^a	25	12
45	MH$^+$–C$_2$H$_5$	1	1	—	—	3	1
44	MH$^+$–C$_2$H$_6$	5	—	100	—	22	11
43	C$_3$H$_7^+$	5	3	3	1	2	2
42	C$_3$H$_6^+$	2	1	4	—	2	7
41	C$_3$H$_5^+$	31	23	23	12	1	—
30	MH$^+$–C$_3$H$_8$	100	52	2	—	14	2
18	NH$_4^+$	60	20	73	16	6	3

^{a} This ion is not seen in H$_2$ CI mass spectra and may arise by reaction of HCO$^+$ with amine, possibly HCO$^+$ + RNH$_2$ → HCONH$_3^+$ + (R – H).

spectra of trimethylamine and five polytertiary amines. In the i-C_4H_{10} CI mass spectra, the MH+ ion was the dominant ion. In the CH_4 CI mass spectrum of $(CH_3)_3N$, MH+ (45%) and [M–H]+ were both abundant. For the polyamines, the [M–H]+ ion varied from 25 to 44% of the total additive ionization and usually was slightly larger than the MH+ ion signal. The CH_4 CI mass spectra of a number of secondary and tertiary amines containing C_{14} to C_{18} n-alkyl groups have been reported;[128] significant MH+ ion signals were observed although [M–H]+ was the base peak in all spectra. In addition, iminium ions resulting nominally from cleavage of the C–C bond alpha to the N were quite pronounced.

Although amines are not amenable to negative ion chemical ionization, biogenic amines have been derivatized with electrophoric agents to give good yields of negative ions indicative of molecular mass by electron capture.[129,130]

XI. NITRO COMPOUNDS

It has been shown[131–133] that nitroaromatic compounds can undergo reduction to the corresponding amine under Brønsted acid CI conditions with H_2, CH_4, i-C_4H_{10}, or NH_3 as reagent gases. Protonation of the amine gives an MH+ ion which is isobaric with the [MH–NO]+ ion which can be formed by fragmentation of the protonated nitro compound. This reduction reaction is favored by high source temperatures and the presence of water in the ion source.[133]

The H_2 and CH_4 CI mass spectra of a selection of substituted nitrobenzenes have been reported[133] where precautions were taken to avoid this reduction process. The greater extent of fragmentation in the H_2 CI mass spectra made these spectra more useful in distinguishing among isomers. The H_2 CI mass spectra of the three nitroanilines are shown in Figure 8. When there is an *ortho* substituent bearing a hydrogen, a significant peak is observed corresponding to [MH–H_2O]+ (Scheme 4). Other fragmentation modes involve loss

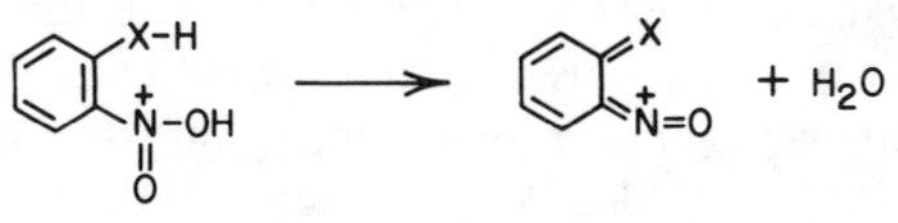

SCHEME 4

of OH, NO, HNO, NO_2, and HNO_2 from the MH+ ion. Loss of OH is particularly prevalent when there is an electron-releasing substituent *ortho* or *para* to the nitro group, while loss of fragments containing NO or NO_2 is particularly pronounced when an electron-releasing substituent is *meta* to the nitro group. Thus, isomeric nitroarenes containing electron-releasing substituents can be identified; identification of isomers containing electron-attracting substituents is much more problematical. The extent of fragmentation of MH+ in the CH_4 CI mass spectra was very small unless elimination of H_2O, as in Scheme 4, was possible.

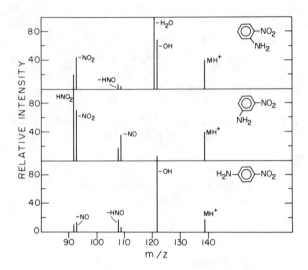

FIGURE 8. H_2 CI mass spectra of nitroanilines. Data from Harrison and Kallury.[133]

The CH_4 and i-C_4H_{10} CI mass spectra of a series of trinitroaromatic compounds also have been reported.[134] The base peak normally is MH^+ with minor fragmentation by loss of OH, H_2O, and apparently, NO from MH^+. It has been shown[135] that reduction to an amine also occurs for the trinitroaromatics and the $[MH–NO]^+$ ion intensities reported earlier[134] may reflect, in part, protonation of the amine formed by this reduction. Many polynitro compounds are explosives which are of forensic interest and the chemical ionization mass spectra of a variety of explosives have been reported.[136]

Many nitroaromatic compounds exhibit strong response to the electron capture detector and should be good candidates for ECCI. Ramdahl and Urdal[137] showed that ECCI was a sensitive method for detecting nitrated polycyclic aromatic hydrocarbons; the major ion observed was M^- with some fragmentation to form $[M–16]^-$ and $[M–30]^-$. Stemmler and Hites[138] have reported the ECCI mass spectra of a large number of nitroaromatic compounds. In many cases intense molecular anions were observed. Common fragmentation reactions involved loss of OH if a labile hydrogen was *ortho* to the nitro group or loss of H, an alkyl radical or an acyl radical to give a phenoxide ion. The fragmentation mechanisms were studied in detail using isotopic labelling and accurate mass measurements. The electron capture chemical ionization mass spectra of a variety of explosives also have been reported.[136,139]

The CH_4 CI mass spectra of various mononitro-, gem-dinitro-, and 1,1,1-trinitroalkanes, halogenonitromethanes, tetranitromethane, and phenylnitromethanes have been reported.[140] In contrast to the EI mass spectra, which gave no molecular ions, the CI mass spectra showed MH^+ ions except for the phenylnitromethanes which eliminated HNO_2 from MH^+ to form stable benzylic ions. Fragmentation of MH^+ in other cases was by loss of H_2O, HNO, and HNO_2; the NO_2^+ ion was abundant in the spectra of the trinitroalkanes. More recently, Ballantine et al.[141] have examined the positive and negative ion CI mass spectra of a selection of aliphatic nitro compounds using NH_3, CH_4 and i-C_4H_{10} as reagent gases; the emphasis was on the ability to derive molecular mass information from the spectra and complete mass spectra were not reported. ECCI gave $[M - H]^-$ for primary and secondary compounds but at a reduced sensitivity compared to Brønsted acid CI. In the latter mode, NH_3 CI proved the best at establishing molecular masses through formation of $[M + NH_4]^+$ for all the compounds studied.

XII. HALOGENATED COMPOUNDS

In terms of the discussion of Section II.A, Chapter 4, hydrogen halide (HX) should be readily lost from the RXH^+ ion formed initially in the Brønsted acid chemical ionization of haloalkanes; in addition, further fragmentation of R^+ will occur if the protonation step is sufficiently exothermic. In line with these predictions, no RXH^+ ions are observed in the H_2 CI of simple alkyl chlorides (except CH_3Cl)[142] or in the H_2 CI of pentyl chlorides.[143] The CH_4 CI mass spectra of the cyclohexyl halides show[118] only $[MH-HX]^+$ while, in the H_2 and CH_4 CI of benzyl halides, $[MH-HX]^+$ is the only significant ion signal.[144,145]

When the halogen is attached to an aromatic nucleus, a much more complex chemistry results. The Brønsted acid CI of halobenzenes has been studied in considerable detail;[144–150] valuable mechanistic information also has been derived from radiation chemical[151,152] and ICR[153] studies. Figure 9 shows the H_2 CI mass spectra of the halobenzenes. Appreciable MH^+ ion signals are observed, except for iodobenzene, despite the strongly exothermic protonation reaction. In addition, charge exchange from H_3^+ to form $M^{+\cdot}$ occurs; this is a common reaction mode of H_3^+. The major fragmentation reactions of MH^+ involve loss of a neutral hydrogen halide to form the phenyl cation or loss of a halogen atom to form the benzene molecular ion. This latter reaction is one of the few examples of formation of an odd-electron product in proton transfer chemical ionization. The phenyl cation adds to H_2 to form the protonated benzene ion. The following reaction scheme is indicated where AH is the reagent gas, AH_2^+ the gaseous Brønsted acid, X is a halogen, and R is a substituent on the phenyl ring.

$$AH_2^+ + RC_6H_4X \rightarrow RC_6H_4XH^{+*} + AH \tag{31}$$

$$RC_6H_4XH^{+*} + M \rightarrow RC_6H_4XH^+ + M \tag{32}$$

$$RC_6H_4XH^{+*} \rightarrow RC_6H_5^+ + X^\cdot \tag{33}$$

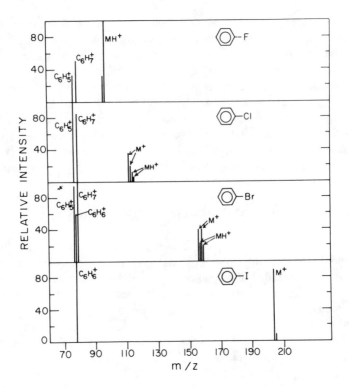

FIGURE 9.   H$_2$ CI mass spectra of halobenzenes. Data from Harrison and Lin[144] and Leung and Harrison.[145]

$$RC_6H_4X \cdot H^{+*} \rightarrow RC_6H_4^+ + HX \tag{34}$$

$$RC_6H_4^+ + AH \rightarrow RC_6H_4A \cdot H^+ \tag{35}$$

In the H$_2$ CI system, the competition between fragmentation by Reactions 33 and 34 is determined primarily by the relative energetics.[145,147] The C$_6$H$_6^{+*}$ ion ($\Delta H_f^0 = 233$ kcal mol^{-1})[154] is more stable than C$_6$H$_5^+$ ($\Delta H_f^0 = 269$ kcal mol^{-1})[154] and tends to drive the reaction to form C$_6$H$_6^{+*}$ + X$\cdot$, with the balance between the two reactions being determined by the relative thermochemical stabilities of HX and X$\cdot$. Table 8 summarizes the heats of formation of the products of Reactions 33 and 34 for the halobenzenes. Because of the thermochemical stability of HF ($\Delta H_f^0 = -65$ kcal mol^{-1}), Reaction 34 is the preferred fragmentation route for C$_6$H$_5$F.H$^+$. As one moves down the halogen family, the stability of HX relative to X$\cdot$ decreases markedly and Reaction 33 becomes competitive with Reaction 34 for protonated bromobenzene while it is the thermochemically favored fragmentation route for protonated iodobenzene.

An electron-releasing substituent, R, stabilizes the RC$_6$H$_5^+$ species more than it stabilizes the RC$_6$H$_4^+$ species. Hence, there is an increased tendency for elimi-

TABLE 8
ΔH_f^0 of Ionic Products in Fragmentation of $C_6H_5X.H^+$

X	$\Delta H_f^0[C_6H_6^+ + X]$ (kcal mol^{-1})	$\Delta H_f^0[C_6H_5^+ + HX]$ (kcal mol^{-1})
F	252	204
Cl	262	247
Br	260	260
I	259	275

nation of X from $RC_6H_4X'H^+$ when electron-releasing substituents are present. Thus, in the H_2 CI mass spectra, both chloroanisoles and chloroanilines show predominant loss of Cl from MH^+, while Br loss is the only significant fragmentation route for protonated bromoanisoles and bromoanilines.[147] When the electron-attracting nitro substituent is present, the fragmentation reactions center around the nitro group rather than the halogen[133] as discussed in Section XI.

Two further features of the H_2 CI mass spectra of haloaromatic compounds deserve note. One is the sequential loss of hydrogen halide from dihalo (and possibly higher) benzenes. For example, the $ClC_6H_4^+$ ion formed from dichlorobenzene or chlorofluorobenzene does not form a stable adduct on reaction with H_2; rather the adduct eliminates HCl to form $C_6H_5^+$, which reacts with H_2 to form $C_6H_7^+$.[150] Finally, the $RC_6H_4^+$ cations frequently show different reactivities towards H_2. Thus, the $[MH - HCl]^+$ ion from 2-chlorobiphenyl is essentially unreactive, that from 3-chlorobiphenyl is very reactive, and that from 4-chlorobiphenyl is moderately reactive towards H_2. As a result, the three isomers can be distinguished by their H_2 CI mass spectra;[155] similar differences are observed for the dichlorobiphenyls but not for the more highly chlorinated species.

The CH_4 CI mass spectra of the halobenzenes and halotoluenes show[145,147] essentially the same features as the H_2 CI mass spectra, however, the extent of fragmentation is much less since the protonation exothermicity is lower. The similarity in behavior includes addition of the phenyl cations to CH_4, Reaction 35. Features specific to the CH_4 CI are the formation of the cluster ions $[M + C_2H_5]^+$ and $[M + C_3H_5]^+$ and the displacement of an iodine atom by $C_2H_5^+$, Reaction 36.

$$C_2H_5^+ + RC_6H_4I \rightarrow C_2H_5C_6H_4R^{+} + I$$ (36)

However, significant differences in the character of the CH_4 and H_2 CI mass spectra are observed for halobenzenes bearing electron-releasing substituents. Table 9 presents the CH_4 CI mass spectra of the isomeric fluoro-, chloro-, and bromo-anilines.[148] The features of note are the almost complete absence of loss of HX from MH^+ for *ortho* substituted compounds and the significant enhancement of this process for *meta* substituted compounds. This is particularly

TABLE 9
CH$_4$ Chemical Ionization Mass Spectra of Haloanilines

Halogen	M.C$_3$H$_5^+$	M.C$_2$H$_5^+$	MH$^+$	M$^+$	[M.C$_2$H$_5^+$-X]	[MH$^+$-X]	[MH$^+$-HX]	[MH$^+$-HX+CH$_4$]
o-F	4.3	11.2	100	8.9	—	—	3.1	—
m-F	9.6	33.6	100	17.2	—	0.9	24.6	22.2
p-F	7.9	27.5	100	28.9	—	5.8	6.7	10.6
o-Cl	7.8	18.6	100	9.3	—	—	—	—
m-Cl	7.1	20.3	100	15.0	—	—	18.6	17.9
p-Cl	7.1	19.6	100	9.7	—	—	1.8	1.8
o-Br	4.7	6.9	100	11.3	18.2	24.3	—	—
m-Br	8.1	4.6	100	16.2	33.2	62.5	10.9	15.9
p-Br	1.6	2.4	58.2	28.2	26.9	100	0.2	—

Relative intensity

noticeable for the chlorobenzene derivatives where $[MH - HCl]^+$ and its adduct to CH_4 are observed only when the chlorine is *meta* to the electron-releasing substituent. This behavior has been observed for a large variety of substituted chlorobenzenes, and formation of $[MH - HCl]^+$ occurs in the CH_4 CI of the *meta* compounds even when formation of $[MH - Cl]^{+\cdot}$ is thermochemically more favorable.[148] These specific substituent effects have been attributed[148] to ion-local bond dipole interactions which influence the extent of protonation at the halogen, although it is clear that there are a variety of subtle structural and energetics effects which also play a role.[150]

There have been numerous investigations of the Brønsted acid chemical ionization masss spectra of polyhalogenated compounds largely as a result of their role as significant environmental contaminants. The majority of these studies have used CH_4 as the reagent gas. The results obtained are, generally, in agreement with the results described above for simpler systems. Thus, the CH_4 CI mass spectra of a series of polycyclic nonaromatic hydrocarbons showed[156,157] very low MH^+ ion signals, the base peak normally being $[MH - HCl]^+$ unless a hydroxyl group was present when $[MH - H_2O]^+$ was the base peak. Further loss of HCl from $[MH - HCl]^+$ was observed as was a retro Diels-Alder reaction. In the same vein, the CH_4 CI mass spectra of the isomeric hexachlorocyclohexanes showed[158] no MH^+ ion signals, the major ion signals corresponding to $[MH - HCl]^+$ and $[MH - 2HCl]^+$. Despite the presence of the basic aromatic system, the CH_4 CI mass spectra of compounds of the type $(ClC_6H_4)_2CHR$ (R = CCl_3, $CHCl_2$, CH_2Cl, CH_3) showed no MH^+ ion signals but rather complete fragmentation to form $[MH - HCl]^+$, $[MH - RH]^+$ and $[MH - ClC_6H_5]^+$, with the latter comprising the base peak.[157]

The CH_4 CI mass spectra of the polybromobenzenes and the polychlorinated benzenes are dominated by MH^+ and the adduct ions $[M + C_2H_5]^+$ and $[M + C_3H_5]^+$.[159] Similar results are obtained for the PCBs.[155,160] In general, it appears that Brønsted acid CI offers little advantage over EI in the identification and quantification of polyhalogenated compounds.

Because of the importance of polyhalogenated hydrocarbons as environmental contaminants, there has been a continuing interest in detecting them at extremely low concentrations; consequently, there have been many studies of the ECCI mass spectra of a variety of halohydrocarbons. The ECCI mass spectra (CH_4 moderating gas) of many halogenated hydrocarbons are included in the compilation by Stemmler and Hites.[161] The same authors also have presented a detailed discussion of the fragmentation reactions of the M^- ions formed in the initial electron capture process.[162] The ECCI mass spectra (CH_4 moderating gas) of a series of bromobenzenes and chlorobenzenes have been reported.[159] For the lower bromobenzenes, Br^- was the dominant product ion; however, for hexabromobenzene, some M^- (~30% of additive ionization) was observed. Similarly, for the lower chlorobenzenes, Cl^- was the dominant ion observed with M^- being dominant (>90% of additive ionization) for pentachloro- and hexachloro-benzene. The same study reported that the ECCI mass spectra of PCBs showed primarily Cl^-. By contrast, 1,2,3,4-tetrachlorodibenzo-*p*-

dioxin showed M⁻, [M – Cl]⁻, and Cl⁻ in its ECCI mass spectrum. Despite the apparent prevalence of formation of Cl⁻ in the ECCI mass spectra of poly-chlorinated compounds, it has been reported[163] that the sensitivity for detection of polychlorodibenzofurans and the higher polychlorodibenzo-*p*-dioxins (the 2,3,7,8-isomer excepted) is orders of magnitude greater by ECCI than by EI. The ECCI mass spectra of polybromobiphenyls show[164] predominantly Br⁻ with minor yields of M⁻ while the ECCI mass spectra of brominated dioxins and dibenzofurans also are dominated[165] by Br⁻.

Crow et al.[159] reported that greater sensitivity for detection of chlorinated compounds was achieved by adding a small amount of oxygen to the methane moderating gas. For the chlorobenzenes, there was an increased abundance of [M – H]⁻ for the lower members of the series and formation of [M – Cl + O]⁻ for the higher members. The displacement of Cl by O⁻ was reported earlier[166] in an atmospheric pressure ionization study of the chemical ionization of polychlorinated compounds using air or N_2 (containing 0.5 ppm O_2) as carrier gas. The phenoxide ions are formed by either or both of the reactions

$$M^- + O_2 \rightarrow [M - Cl + O]^- + OCl^- \tag{37}$$

$$O_2^- + M \rightarrow [M - Cl + O]^- + OCl^- \tag{38}$$

A further reaction which polychlorinated dibenzo-*p*-dioxins undergo is illustrated by Scheme 5. This reaction, which was first observed by Hunt,[167] is

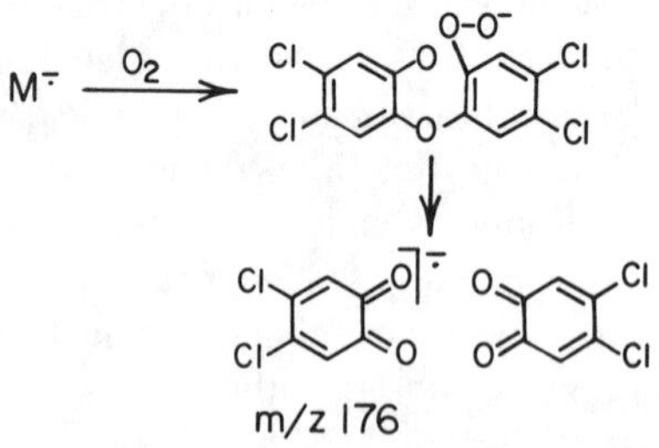

SCHEME 5

particularly prevalent for dioxins not containing a *peri* chlorine, as, for example 2,3,7,8-tetrachlorodibenzo-*p*-dioxin. The identification of polychlorodibenzo-*p*-dioxins by one or both of these reactions involving O_2 has been used extensively[168–170] for identification and quantitation purposes.

When other functional groups are present, further fragmentation routes become possible. Table 10 records the ECCI mass spectra (CH_4 moderating gas) of the trichloroanisoles;[171] losses of H, CH_3, and HCl from the molecular anion are observed in an isomer-dependent fashion. Elimination of H is observed only when there is a H *ortho* to OCH_3, while elimination of HCl is observed

TABLE 10
ECCI Mass Spectra of Trichloroanisoles

Trichloroanisole	Relative intensity		
	[M – H]⁻	[M–CH₃]⁻	[M – HCl]⁻
2,3,4-	30	25	100
2,3,5-	40	30	100
2,3,6-	—	—	100
2,4,5-	5	20	100
2,4,6-	—	—	100
3,4,5-	100	70	—

only when there is a Cl *ortho* to OCH$_3$; isotopic labelling showed that the hydrogen of the HCl originated from the methoxy group. Similar fragmentation reactions involving elimination of HCl and formation of phenoxide ions are observed in the ECCI mass spectra of polychloro-phenoxyanisoles.[172]

The reader is also referred to Section III.A, Chapter 4, for a more general discussion of some of the problems and successes associated with ECCI.

XIII. AMINO ACIDS AND DERIVATIVES

The Brønsted acid CI mass spectra of a variety of amino acids have been determined using H$_2$[173] and CH$_4$[173,174] as reagent gases. An additional study[175] has concentrated on the cluster ions [2M + H]⁺, [M + C$_2$H$_5$]⁺, [M + C$_3$H$_5$]⁺ and their fragmentation modes. The i-C$_4$H$_{10}$ CI mass spectra of a few amino acids also have been reported.[176]

The CH$_4$ CI mass spectra of three representative amino acids are presented in Figure 10. Unsubstituted amino acids, such as leucine, show relatively intense MH⁺ ion signals, the principle fragmentation reaction corresponding to elimination of the elements of formic acid (probably as CO + H$_2$O) from MH⁺ (Reaction 39).

$$RCH(NH_2)CO_2H\cdot H^+ \rightarrow RCH = NH_2^+ + CO_2H_2(CO + H_2O) \quad (39)$$

The weak signal at [MH – 18]⁺ (m/z 114 for leucine) has been shown,[173] using CD$_4$ as reagent gas, to correspond to [M + C$_2$H$_5$ – CO$_2$H$_2$]⁺ rather than [MH – H$_2$O]⁺; the failure to observe [MH – H$_2$O]⁺, the dominant fragment ion for carboxylic acids, has been attributed[173] to a facile decarbonylation of the acyl ion to the very stable RCH=NH$_2$⁺ ion.

$$RCH(NH_2)CO^+ \rightarrow RCH = NH_2^+ + CO \quad (40)$$

When there is a second hydroxyl group present in the molecule, such as for serine, threonine, or aspartic acid (Figure 10), a much more pronounced [MH – 18]⁺ ion signal is observed. It has been shown[173] that the major part of this ion signal

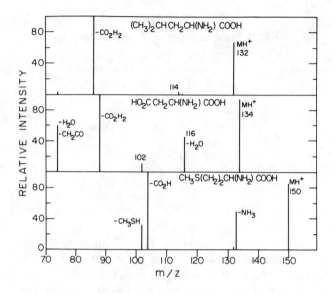

FIGURE 10.   CH$_4$ CI mass spectra of leucine, aspartic acid, and methionine. Data from Tsang and Harrison.

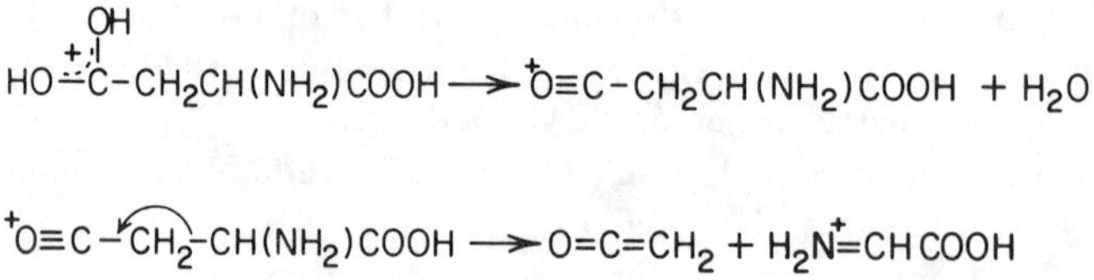

$$HO \overset{\underset{OH}{+\,|\,|}}{=} C - CH_2CH(NH_2)COOH \longrightarrow \overset{+}{O} \equiv C - CH_2CH(NH_2)COOH \ + H_2O$$

$$\overset{+}{O} \equiv C - CH_2 - CH(NH_2)COOH \longrightarrow O = C = CH_2 + H_2\overset{+}{N} = CH\,COOH$$

SCHEME 6

does correspond to H$_2$O elimination from MH$^+$, presumably involving the second hydroxyl function in the molecule as illustrated in Scheme 6 for aspartic acid. As the sequence in the scheme indicates, the [MH – H$_2$O]$^+$ ion of aspartic acid fragments, to a large extent, by elimination of ketene. (The ion signal at m/z 102 for aspartic acid corresponds to elimination of H$_2$O + CH$_2$CO from the [M + C$_2$H$_5$]$^+$ ion.) For serine and threonine, sequential loss of two molecules of H$_2$O from MH$^+$ is observed.

Loss of NH$_3$ from MH$^+$ is observed only when the amino acid contains a functional group which is capable of anchimerically assisting the elimination reaction as shown in Scheme 7 for methionine. Thus, [MH – NH$_3$]$^+$ is observed

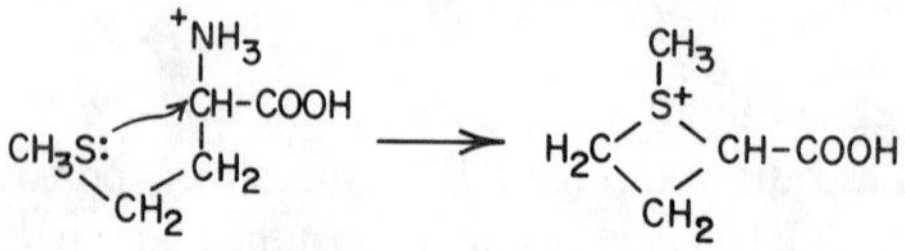

SCHEME 7

in significant abundance only in the Brønsted acid CI mass spectra of such amino acids as methionine, phenylalanine, tyrosine, and tryptophan; indeed, it becomes the base peak in the CH_4 CI mass spectrum of the latter.

The high MH^+ ion intensity in the CH_4 CI mass spectra of amino acids stands in contrast to the electron ionization mass spectra where essentially no molecular ions are observed.[177] In the i-C_4H_{10} CI there is very little fragmentation of the MH^+ ion, although the extent of fragmentation increases markedly with increasing source temperature.[176]

LeClerq and Desiderio[175] have reported the CH_4 CI mass spectra of a number of amino acid amides and methyl esters. Again, abundant MH^+ ion signals are observed with the major fragmentation routes involving loss of NH_3 + CO from the protonated amides and loss of CH_3OH + CO from the protonated methyl esters. Duffield et al., have reported the CH_4 CI mass spectra of a selection of carboxy-*n*-butyl N-trifluoroacetyl[178] and carboxy-*n*-butyl N-penta-flouropropionyl[179] amino acid derivatives. Abundant MH^+ ions were observed, the major fragmentation reactions involving loss of C_4H_8, C_4H_9OH and C_4H_9OH + CO from MH^+. For the diesters of aspartic and glutamic acids, loss of C_4H_8 + C_4H_9OH from MH^+ also were observed while for the hydroxy amino acids, where the OH group also was perfluoroacetylated, loss of $CF_3CO_2H/C_2F_5CO_2H$ was observed as well as loss of the perfluoroacid plus C_4H_8 or C_4H_9OH. The i-C_4H_{10} CI mass spectra of the trifluoroacetyl derivatives showed MH^+ with very little fragmentation.[180]

Smit and Field[15] have reported the OH^- CI mass spectra of two amino acids, leucine and phenylalanine. The $[M – H]^-$ ion dominated the spectra with low-intensity $[M – 3H]^-$ ion signals also being observed. More recently, Vairamani et al.[181] have reported the negative ion mass spectra of a selection of amino acids using Cl^- as reagent ion. For simple amino acids, both $[M – H]^-$ and $[M + Cl]^-$ ions were observed, however, for acids containing an additional polar group, only $[M – H]^-$ ions were observed. Glutamine formed primarily $[M – H – NH_3]^-$ while glutamic acid showed a high yield of $[M – H – H_2O]^-$. The authors reported that the abundances of the $[M – H]^-$ ions were strongly dependent on the source temperature. Thus, alanine, which showed $[M – H]^-$: $[M + Cl]^- = 50:100$ at 200°C source temperature showed no $[M – H]^-$ at 160°C source temperature.

Low and Duffield[179] have reported the ECCI (CH_4 buffer gas) mass spectra of a selection of amino acid carboxy-*n*-butyl N-(O,S)-pentafluoropropionate derivatives. For most compounds studied, $[M – H]^-$ and $[M – HF]^-$ ions, characteristic of the molecular mass, were observed. For aliphatic α-amino acid derivatives, a common fragmentation reaction (Scheme 8) led to $[M – H – C_4H_9CO_2]^-$, which often was the base peak. When the R group of the acid derivative is capable of forming a stable conjugated radical (tyrosine, tryptophan, phenylalanine, and histidine), fragmentation of the $[M – H]^-$ occurred as shown by the second reaction in Scheme 8. Hydroxy amino acids, in which the hydroxyl group also was derivatized, formed predominantly $C_2F_5COO^-$, $[C_2F_5CO_2H–F]^-$, and $C_2F_4COO^-$ rather than ions characteristic of the amino acid. The authors noted that the ECCI mass spectra gave a more

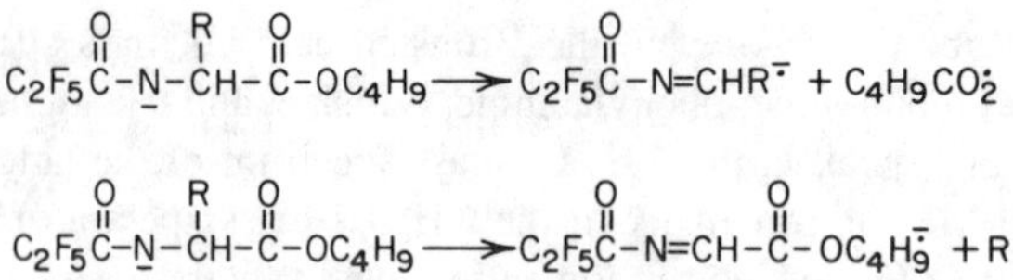

$$
\underset{\substack{\|\\C_2F_5C}}{O} \underset{\substack{|\\H}}{R} \underset{\substack{\|\\C}}{O} - N - CH - C - OC_4H_9 \longrightarrow C_2F_5\overset{O}{\underset{\|}{C}} - N = CHR^- + C_4H_9CO_2^-
$$

$$
\underset{\substack{\|\\C_2F_5C}}{O} \underset{\substack{|\\H}}{R} \underset{\substack{\|\\C}}{O} - N - CH - C - OC_4H_9 \longrightarrow C_2F_5\overset{O}{\underset{\|}{C}} - N = CH - \overset{O}{\underset{\|}{C}} - OC_4H_9^- + R^-
$$

SCHEME 8

ready differentiation of isomers than did the Brønsted acid CI mass spectra and generally showed a greater sensitivity. However, the ECCI mass spectra were strongly dependent on the amount of sample entering the ion source and, therefore, showed considerable variation across the width of the gas chromatographic peak. This could cause serious problems in quantitative work.

The Brønsted acid CI mass spectra of both underivatized and derivatized peptides have been studied extensively using primarily CH_4 and $i\text{-}C_4H_{10}$ as reagent gases.[182–191] Figure 11 illustrates the results obtained in showing the main features of the $i\text{-}C_4H_{10}$ CI mass spectrum of N-acetyl valylalanylalanylphenylalanine methyl ester. MH^+ ions are observed with three series of fragment ions originating by cleavage centered around the amide linkages. The sequence ions are designated in Figure 11 according to the nomenclature proposed by Roepstorff and Fohlman.[192] Clearly, considerable information concerning the amino acid sequence can be derived; collision-induced dissociation of the various fragment ions also has been used to provide sequence information.[189] Nevertheless, there are major problems in vaporizing large fragile peptides for chemical ionization and it is clear that other soft ionization techniques such as plasma desorption and fast atom bombardment combined with tandem mass spectrometry are more suitable for the analysis of larger peptides.[193] A novel recent approach has been the use of laser desorption of the peptide in a FTMS followed by Brønsted acid chemical ionization of the vaporized molecules.[194]

It has been reported that the OH^- CI mass spectra of underivatized peptides show primarily $[M - H]^-$ ions.[195] Sequence information has been obtained from the collision-induced dissociation of these $[M - H]^-$ ions.[195,196]

XIV. CARBOHYDRATES AND DERIVATIVES

The electron ionization mass spectra of carbohydrates and derivatives frequently result from complex fragmentation paths and show very low molecular ion intensities.[197,198] The first indication of the potential of chemical ionization in the analysis of carbohydrates came with the publication of the CH_4 CI mass spectrum of 2-deoxy-D-ribose;[199] although the MH^+ ion intensity was extremely small, a simple fragmentation pattern, resulting from sequential loss of up to three molecules of H_2O from MH^+, was observed. Subsequently, a wide variety of carbohydrates have been examined by chemical ionization using Brønsted acid reagents.[200–206]

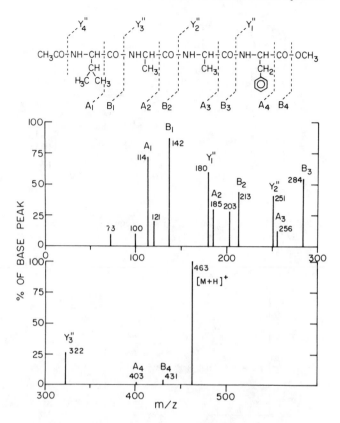

FIGURE 11.   i-C$_4$H$_{10}$ CI mass spectrum of N-acetyl methyl ester deriva-tive of Val-Ala-Ala-Phe. Redrawn from Yu et al.[191]

Chemical ionization using CH$_4$ or i-C$_4$H$_{10}$ as reagent gas results in very low or no MH$^+$ ion signals but does give fragmentation sequences which provide structural information. On the other hand, NH$_3$ CI normally gives molecular weight information through formation of [M + NH$_4$]$^+$ ions. These generalizations are illustrated by the NH$_3$ and i-C$_4$H$_{10}$ CI mass spectra of D-glucopyranose (Figure 12) and methyl β-glucopyranoside tetraacetate (Figure 13). In ammonia CI the [M + NH$_4$]$^+$ ion is the base peak with minor loss of H$_2$O in the case of the D-glucopyranose and loss of [NH$_3$ + CH$_3$OH] from the [M + NH$_4$]$^+$ ion of the tetraacetate. In the i-C$_4$H$_{10}$ CI mass spectra, no MH$^+$ ion is observed but rather a sequence of fragmentation reactions involving the elimination of stable neutral molecules. Of particular note is the initial specific loss of the C1 substituent in the mass spectrum of the tetraacetate. This initial loss of the C1 substituent also has been shown to occur for D-glucopyranose pentaacetate.[200]

Although some sequence information for polysaccharides can be derived from i-C$_4$H$_{10}$ CI mass spectra,[205] NH$_3$ CI with the polysaccharide permethylated

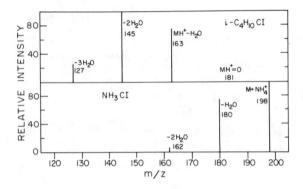

FIGURE 12.   i-C_4H_{10} and NH_3 CI mass spectra of D-glucopyranose. Redrawn from Horton et al.[202]

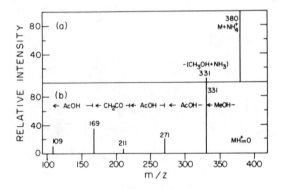

FIGURE 13. NH_3 CI (a) and i-C_4H_{10} CI (b) mass spectra of ß-gluco-pyranose tetraacetate. Data from Horton et al.[202]

and introduced into the source from a direct exposure probe appears to be the most satisfactory approach to sequence determination.[207-209] An example of the spectra obtained is shown in Figure 14 for a reduced permethylated glucoheptasaccharide.[207] Two sequences of fragment ions are observed, as indicated in the figure, each involving cleavage of the glycosidic bond. Further examples of the usefulness of the method are included in the review by Reinhold,[209] who has pointed out that the mass spectra vary with time or the temperature of the heated direct exposure probe. At low temperatures, mostly [M + NH₄]⁺ ions are observed with the fragment ions being observed at higher probe temperatures. Thus, it is not clear whether these fragment ions represent chemical ionization of thermolysis products or direct chemical ionization of intact polysaccharide molecules.

Ganguly et al.[210] have reported the Cl⁻ CI mass spectra of two oligosac-charides. They observed [M + Cl]⁻ as the base peak for both compounds. The

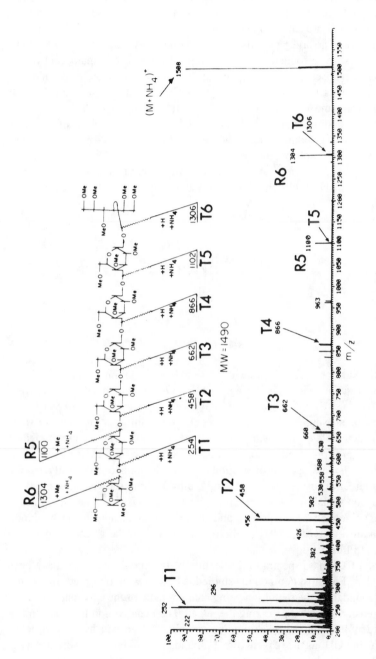

FiGURE 14.   NH₃ DCl mass spectrum of reduced permethylated glucoheptasaccharide. (From Reinhold, V.N. and Carr, S.A., *Anal. Chem.*, 54, 499, 1982. With permission.)

major fragment ions corresponded to the loss of one or more discrete sugar units from either end of the oligosaccharide; all cleavages involved glycosidic linkages and only Cl^- adduct ions were sufficiently intense to be recorded. More recently, Kato and Numajiri[211] have reported the Cl^- chemical ionization mass spectra of a variety of carbohydrates which they obtained using an APCI instrument with introduction of the sample by liquid chromatography. The dominant ion observed in every case was the $[M + Cl]^-$ adduct; in addition, low intensity signals were observed for $[M + Cl + CH_3OH]^-$ with the CH_3OH deriving from the mobile phase of the liquid chromatograph. The Cl^- was produced by adding a small amount of $CHCl_3$ to the mobile phase of the liquid chromatograph. In both studies, it was evident that free hydroxyl groups were necessary for Cl^- attachment to occur.

The i-C_4H_{10} CI mass spectra of several menthyl glucosides/glycosides (Structure A) show substantial MH^+ ion abundances as well as ions resulting

HOCH₂

HO

OH

OR

H

OH

O

OH

A

STRUCTURE A

from cleavage on each side of the glycosidic linkage, giving ions characteristic of both the sugar residue and the aglycone.[212] On the other hand, the i-C_4H_{10}[213] and NH_3[214,215] CI mass spectra of flavanoid glycosides showed very low-intensity or no signals providing molecular mass information, but did show ions characteristic of the sugar and aglycone moieties; even the use of weaker amine reagents provided no ions characteristic of the molecular mass.[215] The ammonia CI mass spectra of a number of cyclopentane iridoid glycosides, using a direct exposure probe for sample introduction, showed substantial $[M + NH_4]^+$ ion signals as well as fragment ions characteristic of the sugar residue and the aglycone.[216] With iridoid and secoiridoid glycosides under CI conditions the use of $(C_2H_5)_2NH_2$ as the reagent gas gave the greatest amount of structural information.[217]

The ECCI mass spectra of a selection of flavanoid glycosides have been reported.[218] For most compounds studied, M^- ions were observed, frequently in low abundance, the major ion being characteristic of the aglycone with only low-intensity ions characterizing the sugar. The negative ion mass spectra of the cardenolide glucosides digitoxin, digoxin, and gitoxin have been determined using OH^- as the reagent ion;[219] abundant $[M - H]^-$ ions were observed as well as ions characteristic of the aglycone and of the sugar sequence. In many of these studies the merit of using a direct exposure probe for introduction of these thermally fragile molecules was clearly demonstrated.

The Brønsted acid CI mass spectra of glucoronides (Structure B) and

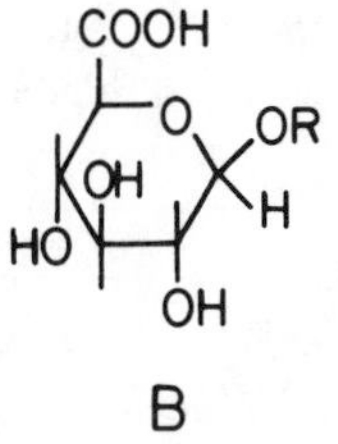

COOH

B

STRUCTURE B

derivatives show very low intensities for ions (MH⁺, [M + NH₄]⁺) from which
the molecular mass can be derived; the fragment ions which are seen derive
primarily from the carbohydrate portion of the molecule,[220–222] although in
some cases, abundant ions characteristic of the aglycone are observed.[223]
Fenselau and co-workers[222,224] have shown that the addition of pyridine to i-
C_4H_{10} or NH_3 gives abundant protonated pyridine ions which form stable
adduct ions with pertrimethylsilyl glucoronide derivatives, thus permitting
molecular masses to be established. Bruins[225] has reported the OH^- CI mass
spectra of a number of underivatized glucoronides using a direct exposure
probe for sample introduction. For most of the samples studied, the $[M - H]^-$
ion was the base peak, with a significant peak corresponding to RO^-. The one
exception occurred when R was *p*-nitrophenyl; in this case, M^- and $[M - OH]^-$
as well as RO^- were observed. It appears in this case that the nitro compound
competed effectively with the reagent gas for capture of the quasi-thermal
electrons. More recently, Brown et al.[226] have shown that the pentafluorobenzyl
esters of pertrimethylsilylated glucoronides show predominantly $[M - C_6F_5CH_2]^-$
in their ECCI mass spectra.

McCloskey and co-workers[227,228] have shown that the CH_4 CI mass spectra of nucleosides (Structure C), consisting of a sugar moiety and a nitrogen

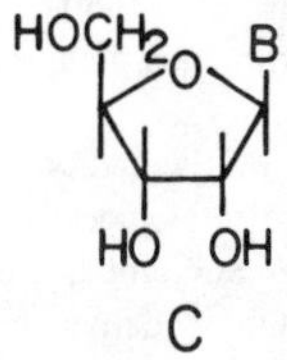

C

STRUCTURE C

base B, show abundant MH⁺ ions, with the fragment ion BH_2^+ usually being
the base peak in the mass spectra. Weaker signals arising from the sugar moiety
are observed. The high MH⁺ abundance is due to the strongly basic nitrogen
centers present in B. Wilson and McCloskey[229] have studied the proton
transfer CI mass spectra of a number of nucleosides using CH_4, i-C_4H_{10}, NH_3,
CH_3NH_2, $(CH_3)_2NH$, and $(CH_3)_3N$ as reagent gases. When proton trans-

fer is not exothermic, adducts of the reagent ion and the nucleoside were observed. From the observance or nonobservance of proton transfer from different reagent ions, approximate PAs of the nucleosides were derived. The mass spectrometry of nucleic acids and related compounds recently has been reviewed in detail.[230]

REFERENCES

1. **Field, F.H., Munson, M.S.B., and Becker, D.A.**, Chemical ionization mass spectrometry, *Adv. Chem. Ser.*, 58, 167, 1966.
2. **Clow, R.P. and Futrell, J.H.**, Ion cyclotron resonance study of the mechanism of chemical ionization. Mass spectroscopy of selected hydrocarbons using methane reagent gas, *J. Am. Chem. Soc.*, 94, 3748, 1972.
3. **Houriet, R., Parisod, G., and Gäumann, T.**, The mechanism of chemical ionization of n-paraffins, *J. Am. Chem. Soc.*, 99, 3599, 1977.
4. **Hiraoka, K. and Kebarle, P.**, Stability and energetics of penta-coordinated carbonium ions. The isomeric $C_2H_7^+$ ions and some higher analogues: $C_3H_9^+$ and $C_4H_{11}^+$, *J. Am. Chem. Soc.*, 98, 6119, 1976.
5. **Hunt, D.F. and McEwen, C.N.**, Chemical ionization mass spectrometry. VII. Deuterium labelled decanes, *Org. Mass Spectrom.*, 7, 441, 1973.
6. **Bowen, R.D. and Williams, D.H.**, The concept of a hierarchy of unimolecular reactions in a homologous series. Prediction of the unimolecular chemistry of some saturated carbenium ions, *J. Chem. Soc. Perkin Trans. 2*, 1479, 1976.
7. **Hunt, D.F. and Harvey, T.M.**, Nitric oxide chemical ionization mass spectra of alkanes, *Anal. Chem.*, 47, 1965, 1975.
8. **Hunt, D.F., McEwen, C.N., and Harvey, T.M.**, Positive and negative chemical ionization mass spectrometry using a Townsend discharge ion source, *Anal. Chem.*, 47, 1730, 1975.
9. **Chai, R. and Harrison, A.G.**, Mixtures of nitric oxide with hydrogen and with methane as chemical ionization gases, *Anal. Chem.*, 55, 969, 1983.
10. **DePuy, C.H., Gronert, S., Barlow, S.E., Bierbaum, V.M., and Damrauer, R.**, The gas-phase acidities of the alkanes, *J. Am. Chem. Soc.*, 111, 1968, 1989.
11. **Field, F.H.**, Chemical ionization mass spectrometry. VIII. Alkenes and alkynes, *J. Am. Chem. Soc.*, 90, 5649, 1968.
12. **Budzikiewicz, H. and Busker, E.**, Studies in chemical ionization mass spectrometry. III. CI-spectra of olefins, *Tetrahedron*, 36, 255, 1980.
13. **Hunt, D.F. and Harvey, T.M.**, Nitric oxide chemical ionization mass spectra of olefins, *Anal. Chem.*, 47, 2136, 1975.
14. **Daishima, S., Iida, Y., and Kanda, F.**, Evaluation of the nitric oxide chemical ionization mass spectra of alcohols and alkenes for isomer distinction, *Anal. Sci.*, 7, 203, 1991.
15. **Smit, A.L.C. and Field, F.H.**, Gaseous anion chemistry. Formation and reactions of OH^-; reactions of anions with N_2O; OH^- negative chemical ionization, *J. Am. Chem. Soc.*, 99, 6471, 1977.
16. **Bruins, A.P.**, Negative ion chemical ionization mass spectrometry in the determination of components in essential oils, *Anal. Chem.*, 51, 967, 1979.
17. **Roy, T.A., Field, F.H., Lin, Y.Y., and Smith, L.L.**, Hydroxyl ion negative chemical ionization mass spectra of steroids, *Anal. Chem.*, 51, 272, 1979.
18. **Bohme, D.K. and Young, L.B.**, Gas phase reaction of oxide radical ion and hydroxide ion with simple olefins and of carbanions with oxygen, *J. Am. Chem. Soc.*, 92, 3301, 1970.
19. **Budzikiewicz, H.**, Structure elucidation by ion-molecule reactions: the location of C,C double and triple bonds, *Fresenius Z. Anal. Chem.*, 321, 150, 1985.

20. **Lankelma, J., Ayanoglu, E., and Djerassi, C.**, Double-bond location in long-chain poly-unsaturated acids by chemical ionization mass spectrometry, *Lipids*, 18, 853, 1983.

21. **Takeuchi, G., Weiss, M., and Harrison, A.G.**, Location of double bonds in alkenyl acetates by negative ion tandem mass spectrometry, *Anal. Chem.*, 59, 918, 1987.

22. **Malosse, C. and Einhorn, J.**, Nitric oxide chemical ionization mass spectrometry of long-chain unsaturated alcohols, acetates, and aldehydes, *Anal. Chem.*, 62, 287, 1990.

23. **Einhorn, J. and Malosse, C.**, Optimized production of the acylium diagnostic ions in the chemical ionization NO$^+$ mass spectra of long-chain monoolefins, *Org. Mass Spectrom.*, 25, 49, 1990.

24. **Budzikiewicz, H., Schneider, B., Busker, E., Boland, W. and Francke, W.**, Studies in chemical ionization mass spectrometry: Part XVI. Are the reactions of aliphatic C:C double bonds with NO$^+$ governed by remote functional groups, *Org. Mass Spectrom.*, 22, 458, 1987.

25. **Schneider, B. and Budzikiewicz, H.**, Experiments on the formation of acylium ions from alkenic compounds following chemical ionization with NO$^+$. *Org. Mass Spectrom.*, 26, 498, 1991.

26. **Peake, D.A. and Gross, M.L.**, Iron(I) chemical ionization and tandem mass spectrometry for locating double bonds, *Anal. Chem.*, 57, 115, 1985.

27. **Field, F.H. and Munson, M.S.B.**, Chemical ionization mass spectrometry. V. Cycloparaffins, *J. Am. Chem. Soc.*, 89, 4272, 1968.

28. **Li, Y.–H., Herman, J.A., and Harrison, A.G.**, Charge exchange mass spectra of some C_5H_{10} isomers, *Can. J. Chem.*, 59, 1753, 1981.

29. **Herman, J.A., Li, Y.–H., and Harrison, A.G.**, Energy dependence of the fragmentation of some isomeric $C_6H_{12}{}^+$ ions, *Org. Mass Spectrom.*, 17, 143, 1982.

30. **Herman, J.A., Li, Y.–H., Lin, M.S., and Harrison, A.G.**, Energy dependence of the fragmentation of the molecular ions of some branched C_7H_{14} olefins, *Int. J. Mass Spectrom. Ion Processes*, 66, 75, 1985.

31. **Busker, E. and Budzikiewicz, H.**, Studies in chemical ionization mass spectrometry 2. i-C_4H_{10} and NO spectra of alkynes, *Org. Mass Spectrom.*, 14, 222, 1979.

32. **Mercer, R.S. and Harrison, A.G.**, Mass spectral behaviour of n-alkynes, *Org. Mass Spectrom.*, 21, 717, 1986.

33. **Munson, M.S.B. and Field, F.H.**, Chemical ionization mass spectrometry. IV. Aromatic hydrocarbons, *J. Am. Chem. Soc.*, 89, 1047, 1967.

34. **Field, F.H.**, Chemical ionization mass spectrometry. VI. C_7H_8 isomers, toluene, cycloheptatriene and norbornadiene, *J. Am. Chem. Soc.*, 89, 5328, 1967.

35. **Leung, H.–W. and Harrison, A.G.**, Hydrogen migrations in mass spectrometry. IV. Formation of $C_6H_7^+$ in the chemical ionization mass spectra of alkylbenzenes, *Org. Mass Spectrom.*, 12, 582, 1977.

36. **Harrison, A.G., Lin, P.–H., and Leung, H.–W.**, The chemical ionization of alkylbenzenes, *Adv. Mass Spectrom.*, 7, 1394, 1978.

37. **Wesdemiotis, C., Schwarz, H., Van de Sande, C.C., and Van Gaever, F.**, Transalkylation and proton-catalyzed C–C bond cleavage of gaseous n-butyl and n-pentyl benzenes, *Z. Naturforsch.*, 34b, 495, 1978.

38. **Herman, J.A. and Harrison, A.G.**, Effect of protonation exothermicity on the chemical ionization mass spectra of some alkylbenzenes, *Org. Mass Spectrom.*, 16, 423, 1981.

39. **Kuck, D.**, Mass spectrometry of alkylbenzenes and related compounds. Part II. Gas phase ion chemistry of protonated alkylbenzenes (alkylbenzenium ions), *Mass Spectrom. Rev.*, 9, 583, 1990

40. **Berthomieu, D., Audier, H., Denhez, J.–P., Monteiro, C., and Mourgues, P.**, [$C_6H_6,C_3H_7^+$] and [$C_6H_7^+,C_3H_6$] ion-neutral complexes intermediate in the reactions of protonated propylbenzenes, *Org. Mass Spectrom.*, 26, 271, 1991.

41. **Hatch, F. and Munson, B.**, Reactant ion monitoring for selective detection in gas chromatography/chemical ionization mass spectrometry, *Anal. Chem.*, 49, 731, 1977.

42. **Hawthorne, S.B. and Miller, D.J.**, Identifying alkylbenzene isomers with chemical ionization-proton exchange mass spectrometry, *Anal. Chem.*, 57, 694, 1985.

43. **Miller, D.J. and Hawthorne, S.B.**, Chemical ionization-proton exchange mass spectrometric identification of alkyl benzenes, tetralins and indans in a coal-derived jet fuel, in *Novel Techniques in Fossil Fuel Mass Spectrometry*, ASTM STP1019, Ashe, T.R. and Wood, K.V., Eds., Am. Soc. Test. Mater., Philadelphia, 1989.

44. **Einolf, N. and Munson, B.**, High pressure charge exchange mass spectrometry, *Int. J. Mass Spectrom. Ion Phys.*, 9, 141, 1972.

45. **Hunt, D.F.**, Reagent gases for chemical ionization mass spectrometry, *Adv. Mass Spectrom.*, 6, 517, 1974.

46. **Bruins, A.P., Ferrer–Correia, A.J., Harrison, A.G., Jennings, K.R., and Mitchum, R.K.**, Negative ion chemical ionization mass spectra of some aromatic compounds using O⁻ as the reagent ion, *Adv. Mass Spectrom.*, 7, 355, 1978.

47. **Jennings, K.R.**, Investigations of selective reagent ions in chemical ionization mass spectrometry, in *High Performance Mass Spectrometry*, Gross, M.L., Ed., American Chemical Society, Washington, 1978.

48. **Jennings, K.R.**, Negative chemical ionization mass spectrometry, in *Mass Spectrometry, Specialist Periodical Report*, Vol. 4, Chemical Society, London, 1977.

49. **Harrison, A.G. and Tong, H.–Y.**, Characterization of C_8H_{10} alkylbenzenes by negative ion mass spectrometry, *Org. Mass Spectrom.*, 23, 135, 1988.

50. **Brotherton, S.A. and Gulick, Jr., W.M.**, Positive and negative-ion chemical ionization gas chromatography/mass spectrometry of polynuclear aromatic hydrocarbons, *Anal. Chim. Acta*, 186, 101, 1986.

51. **Hazell, S.J., Bowen, R.D., and Jennings, K.R.**, The origin of $M^{+\cdot}$ in the chemical ionization mass spectra of polymethylbenzenes, *Org. Mass Spectrom.*, 23, 597, 1988.

52. **Simonsick, Jr., W.J., and Hites, R.A.**, Analysis of isomeric polycyclic aromatic hydrocarbons by charge exchange chemical ionization mass spectrometry, *Anal. Chem.*, 56, 2949, 1984.

53. **Iida, Y. and Daishima, S.**, Negative ion chemical ionization mass spectra of polycyclic aromatic hydrocarbons, *Chem. Lett.*, 273, 1983.

54. **Oehme, M.**, Determination of isomeric polycyclic aromatic hydrocarbons in air particulate matter by high resolution gas chromatography/negative ion chemical ionization mass spectrometry, *Anal. Chem.*, 55, 2290, 1983.

55. **Buchanan, M.J. and Olerich, G.**, Differentiation of polycyclic aromatic hydrocarbons using electron capture negative chemical ionization, *Org. Mass Spectrom.*, 19, 486, 1984.

56. **Field, F.H.**, Chemical ionization mass spectrometry. XII. Alcohols, *J. Am. Chem. Soc.*, 92, 2672, 1970.

57. **Sarris, J., Etievant, P.X., Le Querre, J.–L., and Adda, J.**, The chemical ionization mass spectra of alcohols, in *Progress in Flavour Research 1984*, Adda, J., Ed., Elsevier, Amsterdam, 1985.

58. **Winkler, F.J., Gülaçar, F.O., Mermoud, F., and Buchs, A.**, Temperature dependence of the isobutane chemical ionization of open-chain and cyclic alcohols; structural, stereochemical and molecular-size effects. A reevaluation of the chemical ionization of alcohols, *Helv. Chim. Acta*, 66, 929, 1983.

59. **Gülaçar, F.O., Mermoud, F., Winkler, F.J., and Buchs, A.**, Ammonia chemical ionization mass spectrometry of alcohols; structural, stereochemical, molecular-size and temperature effects, *Helv. Chim. Acta*, 67, 488, 1984.

60. **Munson, B., Feng, T.–M., Ward, H.D., and Murray, Jr., R.K.**, Isobutane chemical ionization mass spectra of unsaturated alcohols, *Org. Mass Spectrom.*, 22, 606, 1987.

61. **Ichikawa, H. and Harrison, A.G.**, Hydrogen migrations in mass spectrometry. VI. The chemical ionization mass spectra of benzoic acids and benzyl alcohols, *Org. Mass Spectrom.*, 13, 389, 1978.

62. **Herman, J.A. and Harrison, A.G.**, Effect of reaction exothermicity on the proton transfer chemical ionization mass spectra of C_5 and C_6 alcohols, *Can. J. Chem.*, 59, 2125, 1981.

63. **Lange, G. and Schultze, W.**, Applications of isobutane and ammonia chemical ionization for the analysis of volatile terpene alcohols and esters, *Flavour and Fragrance Journal*, 2, 63, 1987.

64. **Munson, M.S.B. and Field, F.H.**, Chemical ionization mass spectrometry. I. General introduction, *J. Am. Chem. Soc.*, 88, 2621, 1966.

65. **Dzidic, I. and McCloskey, J.A.**, Influence of remote functional groups on the chemical ionization mass spectra of long-chain compounds, *J. Am. Chem. Soc.*, 93, 4955, 1971.

66. **Winkler, J. and McLafferty, F.W.**, Stereochemical effects in the chemical ionization of cyclic diols, *Tetrahedron*, 30, 2971, 1974.

67. **Munson, B.**, Chemical ionization mass spectrometry: analytical applications of ion-molecule reactions, in *Interactions Between Ions and Molecules*, Ausloos, P., Ed., Plenum Press, New York, 1975.

68. **Lange, G. and Schultze, W.**, Differentiation of isopulegol isomers by chemical ionization mass spectrometry, in *Bioflavours '87*, Schreier, P., Ed., Walter de Gruyter, Berlin, 1988.

69. **Jalonen, J. and Taskinen, J.**, Organic chemistry of gas-phase ions. Part 1. Effect of the protonation site in norborneols, *J. Chem. Soc. Perkin Trans. 2*, 1833, 1985.

70. **Hunt, D.F. and Ryan, J.F.**, Chemical ionization mass spectrometry studies: nitric oxide as a reagent gas, *J. Chem. Soc. Chem. Commun.*, 620, 1972.

71. **Houriet, R., Stahl, D., and Winkler, F.J.**, Negative chemical ionization of alcohols, *Environ. Health Perspect.*, 36, 63, 1980.

72. **Hogge, L.R. and Olson, D.J.H.**, Detection of trace quantities of aliphatic alcohols using derivatization techniques suitable for positive and/or negative ion gas chromatography/chemical ionization mass spectrometry, *J. Chromatog. Sci.*, 20, 109, 1982.

73. **Audier, H.E., Millet, A., Perret, C., and Varenne, P.**, Mecanismes de fragmentation en spectrometrie de masse par ionization chimique. IV. Fragmentation des iones oxiranes protonés, *Org. Mass Spectrom.*, 14, 129, 1973.

74. **Bowen, R.D. and Harrison, A.G.**, Chemical ionization mass spectra of selected C_4H_8O compounds, *J. Chem. Soc. Perkin Trans. 2*, 1544, 1981.

75. **Headley, J.V. and Harrison, A.G.**, Structure and fragmentation of $C_5H_{11}O^+$ ions formed by chemical ionization, *Can. J. Chem.*, 63, 609, 1985.

76. **Fales, H.M., Fenselau, C., and Duncan, J.H.**, The loss of water from isotopically labelled heptanals in chemical ionization, *Org. Mass Spectrom.*, 11, 669, 1976.

77. **Benoit, F.M. and Harrison, A.G.**, Hydrogen migrations in mass spectrometry. I. Loss of olefin from phenyl n-propyl ether following electron impact and chemical ionization, *Org. Mass Spectrom.*, 11, 599, 1976.

78. **Bowen, R.D. and Harrison, A.G.**, Chemical ionization mass spectra of selected C_3H_6O compounds, *Org. Mass Spectrom.*, 16, 159, 1981.

79. **Weiss, M., Crombie, R.A., and Harrison, A.G.**, Structure and fragmentation of $[C_6H_{13}O]^+$ ions formed by chemical ionization, *Org. Mass Spectrom.*, 22, 216, 1987.

80. **Jardine, I. and Fenselau, C.**, The high pressure nitric oxide mass spectra of aldehydes, *Org. Mass Spectrom.*, 10, 748, 1975.

81. **Harrison, A.G. and Jennings, K.R.**, Reactions of O^- with carbonyl compounds, *J. Chem. Soc. Faraday Trans. 1*, 72, 1601, 1976.

82. **Marshall, A., Tkaczyk, M., and Harrison, A.G.**, O^- chemical ionization of carbonyl compounds, *J. Am. Soc. Mass Spectrom.*, 2, 292, 1991.

83. **Dawson, J.H.J., Noest, A.J., and Nibbering, N.M.M.**, 1,1- and 1,3-elimination of water from the reaction complex of O^- with 1,1,1,-trideuteroacetone, *Int. J. Mass Spectrom. Ion Phys.*, 30, 189, 1979.

84. **Weinkam, R.J. and Gal, J.**, Effects of bifunctional interactions in the chemical ionization mass spectrometry of carboxylic acids and methyl esters, *Org. Mass Spectrom.*, 11, 188, 1976.

85. **Harrison, A.G. and Kallury, R.K.M.R.**, Stereochemical applications of mass spectrometry. II. Chemical ionization mass spectra of isomeric dicarboxylic acids and derivatives, *Org. Mass Spectrom.*, 15, 277, 1980.

86. **Benoit, F.M. and Harrison, A.G.**, Predictive value of proton affinity-ionization energy correlations, *J. Am. Chem. Soc.*, 99, 3980, 1977.

87. **Middlemiss, N.E. and Harrison, A.G.**, Structure and fragmentation of gaseous protonated acids, *Can. J. Chem.*, 57, 2827, 1979.

88. **Wienkam, R.J.**, The importance of intramolecular association in the chemical ionization mass spectra of monoenoic and monoepoxy fatty acid methyl esters, *J. Am. Chem. Soc.*, 96, 1032, 1974.

89. **Tsang, C.W. and Harrison, A.G.**, Effect of chain length on the chemical ionization mass spectra of methyl n-alkanoates, *J. Chem. Soc. Perkin Trans. 2*, 1718, 1975.

90. **Harrison, A.G. and Tsang, C.W.**, The hydrogen and methane chemical ionization mass spectra of some formate esters, *Can. J. Chem.*, 54, 2029, 1976.

91. **Herman, J.A. and Harrison, A.G.**, Energetics and structural effects on the fragmentation of protonated esters in the gas phase, *Can. J. Chem.*, 59, 2133, 1981.

92. **Munson, M.S.B. and Field, F.H.**, Chemical ionization mass spectrometry. II. Esters, *J. Am. Chem. Soc.*, 88, 4337, 1966.

93. **May, H.V. and Williams, A.A.**, The chemical ionization of alkyl esters, in *Progress in Flavour Research 1984*, Adda, J., Ed., Elsevier, Amsterdam, 1985.

94. **Plattner, R.D. and Spencer, G.F.**, Chemical ionization mass spectrometry of wax esters, *Lipids*, 18, 68, 1983.

95. **Lau, B.P.–Y., Page, D., and Weber, D.**, Gas chromatography and mass spectrometry of alkyl carbamates, *Can. J. Spectros.*, 34, 53, 1989.

96. **Stamp, J.J., Siegmund, E.G., Cairns, T., and Chan, K.C.**, Chemical ionization of carbamate pesticides; a major dissociation pathway, *Anal. Chem.*, 58, 873, 1986.

97. **Lange, G. and Schultze, W.**, Studies on terpenoid and non-terpenoid esters using chemical ionization mass spectrometry in GC/MS coupling, in *Bioflavours '87*, Schreier, P., Ed., Walter de Grutyer, Berlin, 1988.

98. **Murata, T., Takahashi, S., and Takeda, T.**, Chemical ionization mass spectrometry. II. Applications to analysis of sterol esters, *Anal. Chem.*, 47, 577, 1975.

99. **Lin, Y.Y.**, Identification of steroids by chemical ionization mass spectrometry, *Lipids*, 15, 756, 1980.

100. **Lin, Y.Y. and Smith, L.L.**, Chemical ionization mass spectrometry of steroids and other lipids, *Mass Spectrom. Rev.*, 3, 319, 1984.

101. **Weinkam, R.J. and Gal. J.**, Hydrolysis, methanolysis and ammonolysis of dicarboxylic acids and methyl esters under conditions of chemical ionization, *Org. Mass Spectrom.*, 11, 197, 1976.

102. **Chai, W., Wang, G., Xu, Z., Pan, J., and Duan, S.**, Bifunctional interaction and its effect on the esterification of dicarboxylic acids in chemical ionization mass spectrometry, *Org. Mass Spectrom.*, 18, 64, 1983.

103. **Hawthorne, S.B. and Miller, D.J.**, Water chemical ionization mass spectrometry of aldehydes, ketones, esters and carboxylic acids, *Appl. Spectros.*, 40, 1200, 1986.

104. **Vairamani, M. and Sareswathi, M.**, Negative chemical ionization (Br⁻) mass spectra of dicarboxylic acids, *Org. Mass Spectrom.*, 24, 355, 1989.

105. **Hendricks, H. and Bruins, A.P.**, A tentative identification of components in the essential oil of *Cannabis sativa L.* by a combination of gas chromatography negative ion chemical ionization mass spectrometry and retention indices, *Biomed. Mass Spectrom.*, 10, 377, 1983.

106. **Bambagiotti, M., Giannellini, V., Coran, S.A., Vincieri, F.F., Moneti, G., Selva, A., and Traldi, P.**, Negative ion chemical ionization and collisional activation mass spectrometry of naturally occurring bornyl C_5–C_6 esters, *Biomed. Mass Spectrom.*, 9, 495, 1982.

107. **Bambagiotti, M., Coran, S.A., Giannellini, V., Vincieri, F.F., Daolio, S., and Traldi, P.**, Hydroxyl ion negative ion chemical ionization and collisionally activated dissociation mass analyzed kinetic energy spectra of fatty acid methyl esters, *Org. Mass Spectrom.*, 18, 133, 1983.

108. **Evershed, R.P. and Goad, L.J.**, Capillary gas chromatography/mass spectrometry of cholesteryl esters with negative ammonia chemical ionization, *Biomed. Environ. Mass Spectrom.*, 14, 131, 1987.
109. **Grützmacher, H.–Fr. and Grotemeyer B.**, Fragmentation reactions of some aliphatic esters in the NCI(F⁻) and NCI(NH₂⁻) mass spectra, *Org. Mass Spectrom.*, 19, 135, 1984.
110. **Brumley, W.C. and Sphon, J.A.**, Applications of negative ion chemical ionization, in *Applications of Mass Spectrometry in Food Science*, Gilbert, J., Ed., Elsevier, London, 1987.
111. **Cheung, M., Young, A.B., and Harrison, A.G.**, unpublished results.
112. **Kuksis, A., Marai, L., and Myher, J.J.**, Plasma lipid profiling by liquid chromatography with chloride attachment mass spectrometry, *Lipids*, 26, 240, 1991.
113. **Crawford, C.G. and Plattner, R.D.**, Ammonia chemical ionization mass spectrometry of intact diacyl phosphatidylcholine, *J. Lipid Res.*, 24, 456, 1983.
114. **Ayanoglu, E., Wegmann, A., Pilet, O., Marbury, G.D., Hass, J.R., and Djerassi, C.**, Mass spectrometry of phospholipids. Some applications of desorption chemical ionization and fast atom bombardment, *J. Am. Chem. Soc.*, 106, 5246, 1984.
115. **Stökl, D. and Budzikiewicz, H.**, Negative chemical ionization spectra of aromatic acids and esters or are negative chemical ionization spectra of use only if one knows what one is looking for, *Org. Mass Spectrom.*, 17, 470, 1982.
116. **Chan, C.L. and Brown, C.L.**, Gas phase carboxylate anions from p-nitrobenzyl esters of fatty acids, *Biomed. Mass Spectrom.*, 5, 380, 1978.
117. **Strife, R.J. and Murphy, R.C.**, Preparation of penta-fluorobenzyl esters of arachidonic acid lipoxygenase metabolites; analysis by gas chromatography and negative ion chemical ionization mass spectrometry, *J. Chromatogr.*, 305, 3, 1984.
118. **Jardine, I. and Fenselau, C.**, Proton localization in chemical ionization fragmentation, *J. Am. Chem. Soc.*, 98, 5086, 1976.
119. **Audier, H.E., Millet, A., Perret, C., Tabet, J.–C., and Varenne, P.**, Mecanismes de fragmentation en spectrometrie de masse par ionisation chimique. III. Ions formés par cyclisation, *Org. Mass Spectrom.*, 13, 315, 1978.
120. **Davis, D.V. and Cooks, R.G.**, Site of protonation and bifunctional group interactions in α-ω-hydroxyalkylamines, *Org. Mass Spectrom.*, 16, 176, 1981.
121. **Longevialle, P., Milne, G.W.A., and Fales, H.M.**, Chemical ionization mass spectrometry of complex molecules. XI. Stereochemical and conformational effects in the isobutane chemical ionization mass spectra of some steroidal aminoalcohols, *J. Am. Chem. Soc.*, 95, 6666, 1973.
122. **Longevialle, P., Girard, J.–P., Rossi, J.–C., and Tichy, M.**, Influence of the interfunctional distance on the dehydration of amino alcohols in isobutane chemical ionization mass spectrometry, *Org. Mass Spectrom.*, 14, 414, 1979.
123. **Longevialle, P., Girard, J.–P., Rossi, J.–C. and Tichy, M.**, Isobutane chemical ionization mass spectrometry of ß-amino-alcohols. Evaluation of population of conformers at equilibrium in the gas phase, *Org. Mass Spectrom.*, 15, 268, 1980.
124. **Reiner, E.J ., Poirier, R.A., Peterson, M.R., Csizmadia, I.G., and Harrison, A.G.**, Unimolecular fragmentation of gaseous protonated amines, *Can. J. Chem.*, 64, 1652, 1986.
125. **Reiner, E.J., Bowen, R.D., and Harrison, A.G.**, Collision-induced dissociation mass spectra of protonated amines, *Can. J. Chem.*, 67, 2081, 1989.
126. **Schildcrout, S.M. and Geidner, R.M.**, Comparative chemical ionization mass spectra of tricyclic antidepressant amines and related compounds, *Org. Mass Spectrom.*, 24, 241, 1989.
127. **Whitney, T.A., Klemann, L.P., and Field, F.H.**, Investigation of polytertiary alkylamines using chemical ionization mass spectrometry, *Anal. Chem.*, 43, 1048, 1971.
128. **Valls, M. and Bayona, J.M.**, Characterization of cationic surfactant markers and their abiotic degradation products by CGC-EI/PICI MS, *Fresenius J. Anal. Chem.*, 339, 212, 1991.

129. **Midgley, J.M., MacLachlan, J., and Watson, D.G.,** An extraction-derivatization method suitable for the analysis of biogenic amines by gas chromatography negative ion mass spectrometry, *Biomed. Environ. Mass Spectrom.*, 15, 535, 1988.

130. **Durden, D.A.,** An evaluation of the negative ion mass spectra of electron-capturing derivatives of the biogenic amines. I. Phenylethylamine, *Biol. Mass Spectrom.*, 20, 367, 1991.

131. **Maquestiau, A., Van Haverbeke, Y., Flammang, R., Mispreuve, H., and Elguero, J.,** Ionisation chimique de composés aromatiques nitres, *Org. Mass Spectrom.*, 14, 117, 1979.

132. **Brophy, J.J., Diakiw, V., Goldsack, R.J., Nelson, D., and Shannon, J.S.,** Anomalous ions in the chemical ionization mass spectra of aromatic nitro and nitroso compounds, *Org. Mass Spectrom.*, 14, 201, 1979.

133. **Harrison, A.G. and Kallury, R.K.M.R.,** The chemical ionization mass spectra of mononitroarenes, *Org. Mass Spectrom.*, 15, 284, 1980.

134. **Zitrin, Y. and Yinon, J.,** Chemical ionization mass spectra of 2,4,6-trinitro-aromatic compounds, *Org. Mass Spectrom.*, 11, 388, 1976.

135. **Yinon, J. and Laschever, M.,** Reduction of trinitroaromatic compounds in water by chemical ionization mass spectrometry, *Org. Mass Spectrom.*, 16, 264, 1981.

136. **Yinon, J.,** Mass spectrometry of explosives, in *Forensic Mass Spectrometry*, Yinon, J., Ed., CRC Press, Boca Raton, FL, 1987.

137. **Ramdahl, T. and Urdal, K.,** Determination of nitrated polycyclic aromatic hydrocarbons by fused silica capillary gas chromatography/negative ion chemical ionization mass spectrometry, *Anal. Chem.*, 54, 2256, 1982.

138. **Stemmler, E.A. and Hites, R.A.,** The electron capture negative ion mass spectra of 2,6-dinitroaniline and 2,4-dinitrophenol herbicides and related nitrobenzene derivatives, *Biomed. Environ. Mass Spectrom.*, 14, 417, 1987.

139. **Yinon, J.,** Analysis of explosives by negative ion chemical ionization mass spectrometry, *J. Forensic Sci.*, 25, 401, 1980.

140. **Chizhov, O.S., Kadentsev, V.I., Palmbach, G.G., Burstein, K.I., Shevelev, S.A., and Feinsilberg, A.A.,** Chemical ionization of aliphatic nitro compounds, *Org. Mass Spectrom.*, 13, 611, 1978.

141. **Ballantine, J.A., Barton, J.D., Carter, J.F., Davies, J.P., Smith, K., Stedman, G., and Kingston, E.E,.,** Relative molecular mass information from aliphatic nitro compounds; a chemical ionization study, *Org. Mass Spectrom.*, 23, 1, 1988.

142. **Colosimo, M., Bucci, R., and Brancaleoni, E.,** The stability of alkylchloronium ions during H_3^+-CIMS of simple chloroalkanes, *Int. J. Mass Spectrom. Ion Phys.*, 39, 145, 1981.

143. **Reiner, E.J. and Harrison, A.G.,** Energy redistribution following proton transfer chemical ionization, *Org. Mass Spectrom.*, 19, 343, 1984.

144. **Harrison, A.G. and Lin, P.–H.,** The chemical ionization mass spectra of fluorotoluenes, *Can. J. Chem.*, 53, 1314, 1975.

145. **Leung, H.–W. and Harrison, A.G.,** Structural and energetics effects in the chemical ionization of halogen-substituted benzenes and toluenes, *Can. J. Chem.*, 54, 3439, 1976.

146. **Leung, H.–W., Ichikawa, H., Li, Y.–H., and Harrison, A.G.,** Concerning the mechanism of dehalogenation of halobenzene derivatives by gaseous Brønsted acids, *J. Am. Chem. Soc.*, 100, 2479, 1978.

147. **Leung, H.–W. and Harrison, A.G.,** The role of energetics in the hydrogen chemical ionization of halobenzene derivatives. Estimates of the heats of formation of substituted phenyl cations, *J. Am. Chem. Soc.*, 101, 3168, 1979.

148. **Leung, H.–W. and Harrison, A.G.,** Specific substituent effects in the dehalogenation of halobenzene derivatives by the gaseous Brønsted acid CH_5^+, *J. Am. Chem. Soc.*, 102, 1623, 1980.

149. **Liauw, W.G., Lin, M.S., and Harrison, A.G.,** Effect of protonation exothermicity on the reaction of gaseous Brønsted acids with halobenzene derivatives, *Org. Mass Spectrom.*, 16, 383, 1981.

150. **Liauw, W.G. and Harrison, A.G.,** Site of protonation in the reaction of gaseous Brønsted acids with halobenzene derivatives, *Org. Mass Spectrom.*, 16, 388, 1981.

151. **Cacace, F. and Speranza, M.**, Aromatic substitution in the gas phase. Ambient behaviour of halo- and dihalo-benzenes towards D_2T^+. Tritio-deprotonation and tritio-dehalogenation, *J. Am. Chem. Soc.*, 98, 7299, 1976.

152. **Speranza, M. and Cacace, F.**, Aromatic substitution in the gas phase. On the mechanism of the dehalogenation reactions of halobenzenes and dihalobenzenes promoted by gaseous Brønsted acids, *J. Am. Chem. Soc.*, 99, 3051, 1977.

153. **Speranza, M., Sefcik, M.D., Henis, J.M.S., and Gaspar, P.P.**, Phenylium ($C_6H_5^+$) ion-molecule reactions studied by ion cyclotron resonance spectroscopy, *J. Am. Chem. Soc.*, 99, 5583, 1977.

154. **Lias, S.G., Bartmess, J.E., Liebman, J.F., Holmes, J.L., Levin, R.D., and Mallard, W.G.**, Gas-phase ion and neutral thermochemistry, *J. Phys. Chem. Ref. Data*, 17, Suppl. 1, 1988.

155. **Harrison, A.G., Onuska, F.I., and Tsang, C.W.**, Chemical ionization mass spectra of selected polychlorinated biphenyl isomers, *Anal. Chem.*, 53, 1183, 1981.

156. **Biros, F.J., Dougherty, R.C. and Dalton, J.**, Positive ion chemical ionization mass spectra of polycyclic aromatic pesticides, *Org. Mass Spectrom.*, 6, 1161, 1972.

157. **McKinney, J.D., Oswald, E.O., Palaszek, S.M., and Corbett, B.J.**, Characterization of chlorinated hydrocarbon pesticides of the DDT and polycyclodiene types by electron impact and chemical ionization mass spectrometry, in *Mass Spectrometry and NMR Spectroscopy in Pesticide Chemistry*, Biros, F. and Haque, R., Eds., Plenum Press, New York, 1974.

158. **Oswald, E.O., Albro, P.W., and McKinney, J.D.**, Use of gas-liquid chromatography coupled with chemical ionization and electron impact mass spectrometry for the investigation of potentially hazardous environmental agents and their metabolites, *J. Chromatogr.*, 98, 363, 1974.

159. **Crow, F.W., Bjorseth, A., Knapp, K.T., and Bennett, R.**, Determination of polyhalogenated hydrocarbons by glass capillary gas chromatography-negative ion chemical ionization mass spectrometry, *Anal. Chem.*, 53, 619, 1981.

160. **Cairns, T.C. and Siegmund, E.G.**, Determination of polychlorinated biphenyls by chemical ionization mass spectrometry, *Anal. Chem.*, 53, 1599, 1981.

161. **Stemmler, E.A. and Hites, R.A.**, *Electron Capture Negative Ion Mass Spectra of Environmental Contaminants and Related Compounds*, VCH Publishers, New York, 1988.

162. **Stemmler, E.A. and Hites, R.A.**, The fragmentation of negative ions generated by electron capture: a review with new data, *Biomed. Environ. Mass Spectrom.*, 17, 311, 1988.

163. **Buser, H.–R., Rappe, C., and Berqvist, P.–A.**, Analysis of polychlorinated dibenzofurans and dioxins and related compounds in environmental samples, *Environ. Health Perspect.*, 60, 293, 1985.

164. **Buser, H.–R.**, Selective detection of brominated aromatic compounds using gas chromatography/negative chemical ionization mass spectrometry, *Anal. Chem.*, 58, 2913, 1986.

165. **Donnelly, J.R., Munslow, W.D., Vonnahme, T.L., Nunn, N.J., Hedin, C.M., Sovocool, G.W., and Mitchum, R.K.**, The chemistry and mass spectrometry of brominated dibenzo-p-dioxins and dibenzofurans, *Biomed. Environ. Mass Spectrom.*, 14, 465, 1987.

166. **Dzidic, I., Carroll, D.I., Stillwell, R.N., and Horning, E.C.**, Atmospheric pressure ionization (API) mass spectrometry: formation of phenoxide ions from chlorinated aromatic compounds, *Anal. Chem.*, 47, 1308, 1975.

167. **Hunt, D.F., Harvey, T.M., and Russell, J.W.**, Oxygen as a reagent gas for the analysis of 2,3,7,8-tetrachlorodibenzo-p-dioxin, *J. Chem. Soc. Chem. Commun.*, 151, 1975.

168. **Hass, J.R., Friesen, M.D., and Hoffman, M.K.**, The mass spectrometry of polychlorinated dibenzo-p-dioxins, *Org. Mass Spectrom.*, 14, 9, 1979.

169. **Korfmacher, W.F. and Mitchum, R.K.**, Atmospheric pressure ionization mass spectrometry of the polychlorinated dibenzo-p-dioxins, *Org. Mass Spectrom.*, 19, 299, 1984.

170. **Miles, W.F., Gurprasad, N.P., and Malis, G.P.**, Isomer-specific determination of hexachlorodibenzo-p-dioxins by oxygen negative chemical ionization mass spectrometry, gas chromatography and high-pressure liquid chromatography, *Anal. Chem.*, 57, 1133, 1985.

171. **Busch, K.L., Hass, J.R., and Bursey, M.M.**, The gas enhanced negative ion mass spectra of polychloroanisoles, *Org. Mass Spectrom.*, 13, 604, 1978.

172. **Campbell, J.–A.B., Griffen, D.A., and Deinzer, M.L.**, Electron capture negative ion and positive ion chemical ionization mass spectrometry of polychlorinated phenoxyanisoles, *Org. Mass Spectrom.*, 20, 122, 1985.

173. **Tsang, C.W. and Harrison, A.G.**, The chemical ionization of amino acids, *J. Am. Chem. Soc.*, 98, 1301, 1976.

174. **Milne, G. W.A., Axenrod, T., and Fales, H.M.**, Chemical ionization mass spectrometry of complex molecules. IV. Amino acids, *J. Am. Chem. Soc.*, 92, 5170, 1970.

175. **LeClerq, P.A. and Desiderio, D.M.**, Chemical ionization mass spectra of amino acids and derivatives. Occurrence and fragmentation of ion-molecule reaction products, *Org. Mass Spectrom.*, 7, 515, 1973.

176. **Meot-Ner, M. and Field, F. H.**, Chemical ionization mass spectrometry. XX. Energy effects and virtual ion temperatures in the decomposition of amino acids and amino acid derivatives, *J. Am. Chem. Soc.*, 95, 7207, 1973.

177. **Junk, G. and Svec, H.**, The mass spectra of α-amino acids, *J. Am. Chem. Soc.*, 85, 839, 1963.

178. **Kingston, E.E. and Duffield, A.M.**, Plasma amino acid quantification using gas chromatography chemical ionization mass spectrometry and ^{13}C amino acids as internal standards, *Biomed. Mass Spectrom.*, 5, 621, 1978.

179. **Low, G.K.–C. and Duffield, A.M.**, Positive and negative chemical ionization mass spectra of amino acid carboxy-*n*-butyl ester N-pentafluoropropionate derivatives, *Biomed. Mass Spectrom.*, 11, 223, 1984.

180. **Finlayson, P.J., Christopher, R.K., and Duffield, A.M.**, Quantitation of fourteen urinary α-amino acids using isobutane gas chromatography chemical ionization mass spectrometry with ^{13}C amino acids as internal standards, *Biomed. Mass Spectrom.*, 7, 450, 1981.

181. **Vairamani, M., Srinivas, R., and Viswandha Rao, G.K.**, Negative ion chemical ionization (Cl⁻) mass spectra of amino acids, *Biomed. Environ. Mass Spectrom.*, 17, 299, 1988.

182. **Gray, W.R., Wojcik, L.H., and Futrell, J.H.**, Application of mass spectrometry to protein chemistry. II. Chemical ionization studies on acetylated permethylated peptides, *Biochem. Biophys. Res. Commun.*, 41, 1111, 1970.

183. **Kiryuskin, A.A., Fales, H.M., Axenrod, T., Gilbert, E.J., and Milne, G.W.A.**, Chemical ionization mass spectrometry of complex molecules. VI. Peptides, *Org. Mass Spectrom.*, 5, 19, 1971.

184. **Bowen, D. V. and Field, F. H.**, Isobutane chemical ionization mass spectrometry of dipeptides, *Int. J. Pept. Protein Res.*, 5, 436, 1973.

185. **Baldwin, M. A. and McLafferty, F. W.**, Direct chemical ionization of relatively involatile samples. Application to underivatized oligopeptides, *Org. Mass Spectrom.*, 7, 1353, 1973.

186. **Beuhler, R. J., Flanigan, E., Greene, L.J., and Friedman, L.**, Proton transfer mass spectrometry of peptides. A rapid heating technique for underivatized peptides containing arginine, *J. Am. Chem. Soc.*, 96, 3990, 1974.

187. **Mudgett, M., Bowen, R. V., Kendt, T.J., and Field, F.H.**, C-methylation: an artifact in peptides derivatized for sequencing by mass spectrometry, *Biomed. Mass Spectrom.*, 2, 254, 1975.

188. **Mudgett, M., Bowen, D.V., Field, F.H., and Kendt, T.J.**, Peptide sequencing: the utility of chemical ionization mass spectrometry, *Biomed. Mass Spectrom.*, 4, 159, 1977.

189. **Hunt, D.F., Buko, A.M., Ballard, J.M., Shabanowitz, J., and Giardini, A.B.**, Sequence analysis of polypeptides by collision activated dissociation on a triple quadrupole mass spectrometer, *Biomed. Mass Spectrom.*, 8, 397, 1981.

190. **Yu, T.J., Schwartz, H., Giese, R.W., Karger, B.L., and Vouros, P.**, Analysis of N-acetyl-N,O,S-permethylated peptides by combined liquid chromatography-mass spectrometry, *J. Chromatogr.*, 218, 519, 1981.

191. **Yu, T.J., Schwartz, H., Cohen, S.A., Vouros, P., and Karger, B.L.**, Sequence analysis of peptides by high performance liquid chromatography-mass spectrometry, *J. Chromatogr.*, 301, 425, 1984.

192. **Roepstorff, P. and Fohlman, J.**, Proposal for a common nomenclature for sequence ions in mass spectra of peptides, *Biomed. Mass Spectrom.*, 11, 601, 1984.
193. **Desiderio, D.M.**, Ed., *Mass Spectrometry of Peptides*, CRC Press, Boca Raton, FL, 1991.
194. **Speir, J.P., Gorman, G.S., Cornett, D.S., and Amster, I.J.**, Controlling the dissociation of peptide ions using laser desorption/chemical ionization Fourier transform mass spectrometry, *Anal. Chem.*, 63, 55, 1991.
195. **Bradley, C.V., Howe, I., and Beynon, J.H.**, Sequence analysis of underivatized peptides by negative ion chemical ionization and collision-induced dissociation, *Biomed. Mass Spectrom.*, 8, 85, 1981.
196. **Eckersley, M., Bowie, J.H., and Hayes, R.N.**, Collision-induced dissociation of deprotonated peptides, dipeptides and tripeptides with hydrogen and alkyl α groups, *Org. Mass Spectrom.*, 24, 597, 1989.
197. **Radford, T. and DeJongh, D.C.**, Carbohydrates, in *Biochemical Applications of Mass Spectrometry*, Waller, G.R., Ed., Wiley-Interscience, New York, 1972.
198. **Radford, T. and DeJongh, D.C.**, Carbohydrates, in *Biochemical Applications of Mass Spectrometry, First Supplementary Volume*, Waller, G.R. and Dermer, O.C., Eds., Wiley-Interscience, New York, 1980.
199. **Fales, H.M., Milne, G.W.A., and Vestal, M.L.**, Chemical ionization mass spectrometry of complex molecules, *J. Am. Chem. Soc.*, 91, 3682, 1969.
200. **Hogg, A.M. and Nagabhusan, T.L.**, Chemical ionization mass spectra of sugars, *Tetrahedron Lett.*, 4827, 1972.
201. **Dougherty, R.C., Horton, D., Philips, K.D., and Wander, J.D.**, The high resolution mass spectrum of 2-acetamido-1,3,4,6-tetra-O-acetyl-2-dioxy-α-D-glucopyranose, *Org. Mass Spectrom.*, 7, 805, 1973.
202. **Horton, D. B., Wander, J. D., and Foltz, R. L.**, Analysis of sugar derivatives by chemical ionization mass spectrometry, *Carbohyd. Res.*, 36, 75, 1974.
203. **Dougherty, R.C., Roberts, J.D., Binkley, W.W., Chizhov, O.S., Kadentsev, V.I., and Solov'yov, A.A.**, Ammonia-isobutane chemical ionization mass spectra of oligosaccharide peracetates, *J. Org. Chem.*, 39, 451, 1974.
204. **Chizhov, O.S., Kadentsev, V.I., Solov'yov, A.A., Brinkley, W.W., Roberts, J.D., and Dougherty, R. C.**, Oligosaccharide acetate mass spectra obtained by addition of ammonium ions, *Dokl. Akad. Nauk. SSSR* (English transl.), 217, 511, 1974.
205. **Chizhov, O.S., Kadentsev, V.I., Solov'yov, A.A., Levonowich, P.F., and Dougherty, R.C.**, Polysaccharide sequencing by mass spectrometry. Chemical ionization spectra of permethylglycosylalditols, *J. Org. Chem.*, 41, 3425, 1976.
206. **Hedin, P.A. and Phillips, V.A.**, Chemical ionization (methane) mass spectrometry of sugars and their derivatives, *J. Agric. Food Chem.*, 39, 1106, 1991.
207. **Reinhold, V.N. and Carr, S.A.**, Direct chemical ionization mass spectrometry with polyimide-coated wires, *Anal. Chem.*, 54, 499, 1982.
208. **Reinhold, V.N. and Carr, S.A.**, New mass spectral approaches to complex carbohydrate structure, *Mass Spectrom. Rev.*, 2, 153, 1983.
209. **Reinhold, V.N.**, Direct chemical ionization mass spectrometry of carbohydrates, in *Complex Carbohydrates, Part E, Methods in Enzymology*, Vol. 138, Ginsburg, V., Ed., Academic Press, Orlando, 1987.
210. **Ganguly, A.K., Cappuccino, N.F., Fujiwara, H., and Bose, A.K.**, Convenient mass spectral technique for structural studies in oligosaccharides, *J. Chem. Soc. Chem. Commun.*, 148, 1979.
211. **Kato, Y. and Numajiri, Y.**, Chloride attachment negative-ion mass spectra of sugars by combined liquid chromatography and atmospheric pressure chemical ionization mass spectrometry, *J. Chromatogr.*, 562, 81, 1991.
212. **Takeda, N., Harada, K., Suzuki, M., Tatematsu, A., and Sakata, I.**, Structural characterization of underivatized menthyl glycosides using chemical ionization mass spectrometry, *Biomed. Mass Spectrom.*, 10, 608, 1983.

213. **Itokawa, H., Oshida, Y., Ikuta, A., and Shida, Y.**, In-beam electron impact, chemical ionization and negative ion chemical ionization of flavanoid glycosides, *Chem. Lett.*, 49, 1982.

214. **Caccamese, S., Amico, V., and Hardy, M.**, Desorption/chemical ionization mass spectra of some underivatized flavanoid glycosides, *Bull. Soc. Chim. France*, 91, 1988.

215. **Mollova, N.N., Bankova, V.S., and Popov, S.S.**, Chemical ionization mass spectrometry with amines as reactant gases. V. Amine chemical ionization mass spectrometry of flavanoid glycosides, *Org. Mass Spectrom.*, 22, 334, 1987.

216. **Demirev, P.A., Handjieva, N., Saadi, H, Popov, S.S., Reshetova, O.S., and Rozynov, B.V.**, Ammonia/desorption chemical ionization of some cyclopentane iridoid glycosides, *Org. Mass Spectrom.*, 26, 151, 1991.

217. **Mollova, N.N., Handjieva, N.V., and Popov, S.S.**, Chemical ionization with amines as reagent gases. VII. Amine chemical ionization mass spectrometry of some iridoid and secoiridoid glucosides, *Org. Mass Spectrom.*, 24, 1001, 1989.

218. **Sakushima, A., Nishibe, S., and Brandenberger, H.**, Negative ion desorption chemical ionization mass spectrometry of flavanoid glycosides, *Biomed. Environ. Mass Spectrom.*, 18, 809, 1989.

219. **Bruins, A.P.**, Negative ion desorption chemical ionization mass spectrometry of digitoxin and related cardenolides, *Int. J. Mass Spectrom. Ion Phys.*, 48, 185, 1983.

220. **Heaney Kieras, J., Kieras, F., and Bowen, D.**, 2-O-methyl-D-glucoronic acid, a new hexuronic acid of biological origin, *Biochem. J.*, 155, 181, 1976.

221. **Lyle, M., Pallante, S., Head, K., and Fenselau, C.**, Synthesis and characterization of glucoronides of cannabinol, cannabidiol, Δ^9-tetrahydrocannabinol and Δ^8-tetrahydrocannabinol, *Biomed. Mass Spectrom.*, 4, 190, 1977.

222. **Johnson, L.P., Subba Rao, S.C., and Fenselau, C.**, Pyridine as a reagent gas for the characterization of glucoronides by chemical ionization mass spectrometry, *Anal. Chem.*, 50, 2022, 1978.

223. **Cairns, T. and Siegmund, E.G.**, Characterization of glucoronides by chemical ionization mass spectrometry with ammonia as a reagent gas, *Anal. Chem.*, 54, 2456, 1982.

224. **Fenselau, C., Cotter, R., and Johnson, L.**, Mass spectral techniques for the analysis of glucoronides, *Adv. Mass Spectrom.*, 8, 1159, 1980.

225. **Bruins, A.P.**, Negative ion desorption chemical ionization mass spectrometry of some underivatized glucoronides, *Biomed. Mass Spectrom.*, 8, 31, 1981.

226. **Brown, S.Y., Garland, W.A., and Fukuda, E.K.**, Gas chromatography/negative chemical ionization mass spectrometry of intact glucoronides, *Biomed. Environ. Mass Spectrom.*, 19, 32, 1990.

227. **Wilson, M.S., Dzidic, I., and McCloskey, J.A.**, Chemical ionization mass spectrometry of nucleosides, *Biochem. Biophys. Acta*, 240, 623, 1971.

228. **McCloskey, J.A., Futrell, J.H., Elwood, T.A., Schram, K.H., Panzica, R.P., and Townsend, L.B.**, Determination of relative glycosyl bond strengths in nucleosides by chemical ionization mass spectrometry. A comparative study of 7- and 9-ß-D-ribofuranosylpurines, *J. Am. Chem. Soc.*, 95, 5762, 1973.

229. **Wilson, M.S. and McCloskey, J.A.**, Chemical ionization mass spectra of nucleosides: mechanism of ion formation and estimation of proton affinity, *J. Am. Chem. Soc.*, 97, 3436, 1975.

230. **Crain, P.F.**, Mass spectrometric techniques in nucleic acid research, *Mass Spectrom. Rev.*, 9, 505, 1990.

Chapter 6

SELECTED TOPICS IN CIMS

I. INTRODUCTION

This chapter reviews four specific topics; isotope exchange reactions in CI systems, stereochemical effects in chemical ionization, tandem mass spectrometry combined with chemical ionization, and reactive collisions in quadrupole collision cells. These topics have not been covered in detail in earlier chapters.

II. ISOTOPE EXCHANGE REACTIONS IN CHEMICAL IONIZATION STUDIES

The exchange of hydrogen for deuterium in organic molecules has found wide use in structural studies in mass spectrometry[1,2] and in mechanistic studies in gas-phase ion chemistry.[3] Normally, the exchange is carried out in solution reactions before the sample is introduced into the mass spectrometer, although in some cases, the exchange reaction can be carried out in the inlet system of the mass spectrometer. There now is very extensive evidence that, at the pressures applicable in chemical ionization, exchange of hydrogen for deuterium can occur in the ion source of the mass spectrometer.

The simplest of these isotopic exchange processes involves the exchange of active hydrogens with a suitable reagent gas. Hunt et al.[4] first showed that the hydrogens bonded to heteroatoms in alcohols, phenols, carboxylic acids, amines, amides, and thiols underwent rapid exchange for deuterium under Brønsted acid CI conditions using D_2O as the reagent gas. The mass shift on substituting D_2O for H_2O as the reagent gas thus provides a count of the number of such hydrogens in the sample molecule. In a similar fashion, the use of ND_3^5 or CH_3OD^6 as reagent gas permits the differentiation of primary, secondary, and tertiary amines from the mass shift when the deuterated reagent gas is substituted for the undeuterated reagent gas. Lin and Smith[7] have shown that the use of ND_3 as reagent gas allows the determination of the number of active hydrogens in a variety of natural products; again a comparison is made of the mass of the protonated ion observed with the undeuterated reagent gas, with the mass of the MD+ ion observed with the deuterated reagent gas. With ND_3 as reagent gas isotopic exchange of active hydrogens has been used to aid in the characterization of nitrogen-containing compounds in the alkaline fractions of a coal-derived liquid,[8] and exchange of active hydrogens with CH_3OD has been used to aid in the characterization of oxygen-containing aromatics in a neutral polar subfraction of a coal-derived liquid.[9] Exchange reactions also have been used to identify the number of active hydrogens in morphine alkaloids[10] and to aid in the identification of heteroaromatic compounds in coal liquefaction and gasification products.[11]

Exchange of aromatic hydrogens in the gas phase was observed first by Freiser et al.[12] Using ICR techniques, they observed sequential replacement of hydrogen by deuterium when protonated benzene reacted with D_2O. A number of fluorine- and methyl-substituted benzenes also were observed to exchange with D_2O although the rate and extent of exchange was found to be very structure-dependent. Thus, although *o*- and *p*-xylene exchanged all aromatic hydrogens rapidly, *m*-xylene showed exchange of one hydrogen only. No exchange was observed for benzenes containing strong electron-releasing or electron-withdrawing substituents. Many of these species protonate on the substituent and it was concluded that protonation of the aromatic ring was a prerequisite for exchange of the aromatic hydrogens to occur. Martinson and Butrill[13] recorded similar observations and reached similar conclusions in a study of the Brønsted acid (D_2O) chemical ionization mass spectra of a variety of aromatic compounds.

An extensive study of the exchange of hydrogen for deuterium in a variety of compounds using both Brønsted acid and Brønsted base CI has been reported by Hunt and Sethi.[14] In positive ion studies using D_2O as reagent gas, they observed that the dominant ion in the CI mass spectra of benzene, toluene, *o*-xylene, and *p*-xylene corresponded to the MD$^+$ ion in which all aromatic hydrogens had been replaced by deuterium. Slow exchange of hydrogens was observed for *m*-xylene and mesitylene, while no exchange of aromatic hydrogens was observed for aromatic compounds containing amine, hydroxy, alkoxy, acetyl, or nitrile substituents, for polycyclic aromatic hydrocarbons or for the metallocenes investigated. Exchange of active hydrogens did occur.

When the more basic reagent gas C_2H_5OD was used, the rate of exchange of the aromatic hydrogens in *m*-xylene and mesitylene increased substantially, the most prominent ion corresponding to MD$^+$ where all aromatic hydrogens had been exchanged. The exchange rate for the oxygen-containing derivatives also increased although there was some evidence that those hydrogens *ortho* or *para* to the substituent exchanged more rapidly than those *meta* to the substituent. In addition, exchange of hydrogens in polycyclic aromatic hydrocarbons, except pyrene, occurred with C_2H_5OD as reagent gas. Replacement of aromatic hydrogens was not observed for aminobenzenes, benzonitrile, acetophenone, or ferrocene.

More recently, Hawthorne and Miller have studied the hydrogen/deuterium exchange reactions of alkylbenzenes[15] and alkyl-tetralins and indans[16] using CH_3OD as the reagent gas. They found that the number of alkyl substituents on the benzene ring could be determined from the extent of deuterium incorporation and that *meta* disubstituted compounds could be distinguished from the *ortho* and *para* isomers because of the lower extent of incorporation of deuterium in the *meta* isomer. For the alkyltetralins and indans only the hydrogens on the aromatic ring underwent exchange; consequently, it was possible to distinguish between substitution on the aromatic ring and substitution on the saturated ring as well as to establish the number of substituents on the aromatic ring. Typical results they have reported[15] for C_9 alkylbenzenes are shown in Figure 1. Note in particular the smaller extent of exchange for *m*-ethyltoluene compared to the *ortho* and *para* isomers as well as the low extent of exchange of 1,3,5-trimethyltoluene (mesitylene). In this regard CH_3OD behaves more like D_2O than C_2H_5OD; this aspect will be discussed in further detail below.

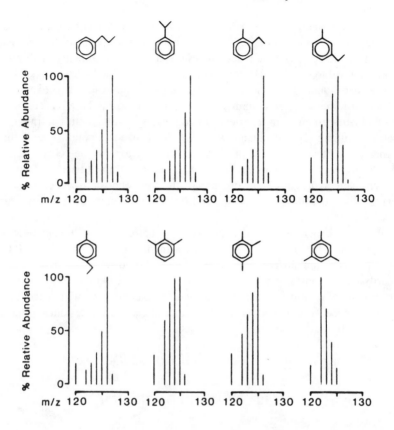

FIGURE 1. Chemical ionization-proton exchange mass spectra of C_9 alkylbenzenes with CH_3OD reagent gas. (From Hawthorne, S.B. and Miller, D.J., *Anal. Chem.*, 57, 694, 1985. With permission.)

Hunt and Sethi[14] have studied the exchange reactions of a few compounds using the very basic reagent ND_3. Slow exchange of the aromatic hydrogens in *m*-toluidine, *m*-phenylenediamine, ferrocene, and vinylferrocene was observed. Aniline, *o*- and *p*-toluidine, *o*- and *p*-phenylenediamine, ruthenocene, hydroxymethylferrocene, and ferrocene carboxaldehyde did not exchange aromatic hydrogens with any of the deuterated reagent gases.

In studies involving negative ions, Stewart et al.[17] have shown that the $[M - H]^-$ ions from simple esters, olefins, acetylenes, allene, and toluene undergo isotopic exchange when allowed to react with D_2O; these experiments were performed in a flowing afterglow apparatus. In a later study[18] the $[M - H]^-$ ions from ketones and aldehydes were shown to exchange with CH_3OD under flowing afterglow conditions. Negative ion studies under chemical ionization conditions were carried out by Hunt and Sethi[14] using OD^- (in D_2O), $C_2H_5O^-$ (in C_2H_5OD), and ND_2^- (in ND_3) as reagents. When OD^- was used as a reagent ion, the $[M - H]^-$ ions of alkylbenzenes showed exchange of all benzylic hydrogens and slow exchange was observed for hydrogen atoms adjacent to the carbonyl group of simple ketones. The weaker base $C_2H_5O^-$ did not abstract a proton from alkylbenzenes but did promote the exchange reaction in simple ketones, aldehydes, and esters; the most prominent ion observed corresponded to the $[M - H]^-$ ion where all the enolizable hydrogens had been replaced by deuterium. Incomplete exchange of enolizable hydrogens was observed for conjugated carbonyl compounds. The strong base ND_2^- abstracts an aromatic proton from a variety of compounds with the resulting $[M - H]^-$ ion undergoing rapid exchange with ND_3 for benzene, naphthalene, anthracene, and phenanthrene, with slower exchange for pyrene, chrysene and benz[*a*]pyrene. The rate of exchange also was slow for toluene. Exchange of the aromatic hydrogens in the organometallic compounds ferrocene, ferrocene carboxaldehyde, phenyl ferrocene, and (toluene) chromium tricarbonyl also was observed.

The exchange reaction is believed to be the result of collisions between sample ions (HR^-, H_2RD^+) and reagent gas neutrals $R'OD$. At 0.5 to 1.0 torr pressure, each ion suffers up to several hundred collisions with neutrals (primarily the reagent gas) before exiting the ion source. On the other hand, the probability of multiple collisions of a given sample molecule with reagent ions is very low because of the low concentrations of each; this rules out significant exchange by a reaction sequence such as

$$H_2R + R'OD_2^+ \rightarrow H_2RD^+ + R'OD \tag{1}$$

$$H_2RD^+R'OD \rightarrow HRD + R'ODH^+ \tag{2}$$

$$HRD + R'OD_2^+ \rightarrow HRD_2^+ + R'OD \tag{3}$$

Rather, the incorporation of deuterium into the sample ion must occur during the lifetime of the ion/molecule collision complex formed when the sample ion and the deuterium-labelled reagent gas collide.

There is substantial evidence[19–20] that the potential energy profile for proton transfer between an ion and a neutral contains two minima with an intermediate potential energy barrier, as illustrated schematically in Figure 2 for the case of negative ions. The two minima correspond to $R'O^-$ hydrogen-bonded to HRH and HR^- hydrogen-bonded to $R'OH$ (or isotopic variants). The exchange reaction occurs when HR^- reacts with the reagent gas $R'OD$ (right to left of Figure 2) and the system has sufficient energy to overcome the intermediate energy barrier (as in the lower curve) to form $HR–D\cdots^-OR'$ with an energy part-way up the curve leading to $HRD + {}^-OR'$. At this point, the hydrogen bonding is weak permitting rotation of HRD leading to $DR–H\cdots^-OR'$, which exits to the right to give $DR^- + R'OH$ as summarized in Scheme 1. Repeated collision with new $R'OD$ reagent molecules leads to multiple D incorporation in the sample ion.

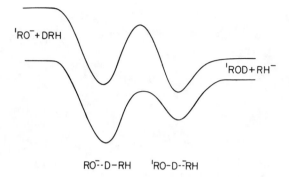

FIGURE 2. Schematic potential energy profile for reaction of RO⁻ with DRH.

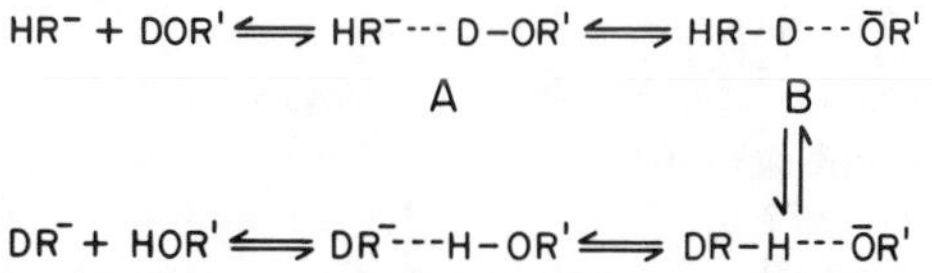

$$HR^- + DOR' \rightleftharpoons HR^- \cdots D{-}OR' \rightleftharpoons HR{-}D \cdots \bar{O}R'$$

$$\qquad\qquad\qquad\qquad A \qquad\qquad\qquad\qquad B$$

$$\qquad\qquad\qquad\qquad\qquad\qquad\qquad\qquad \Updownarrow$$

$$DR^- + HOR' \rightleftharpoons DR^- \cdots H{-}OR' \rightleftharpoons DR{-}H \cdots \bar{O}R'$$

SCHEME 1

$$H_2RD^+ + DOR' \rightleftharpoons DHR^+H \cdots OR'_{|D} \rightleftharpoons DHR \cdots H{-}\overset{+}{O}R'_{|D}$$

$$\qquad\qquad\qquad\qquad\qquad\qquad \Updownarrow$$

$$DHRD^+ + HOR' \rightleftharpoons DHR^+_- D \cdots OR'_{|H} \rightleftharpoons DHR \cdots D{-}\overset{+}{O}R_{|H}$$

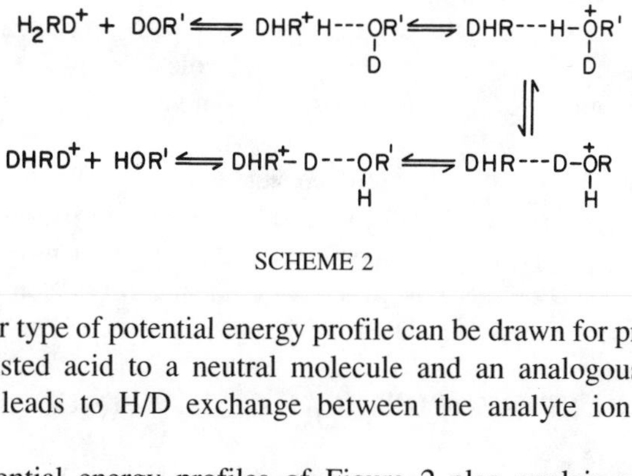

SCHEME 2

A similar type of potential energy profile can be drawn for proton transfer from a Brønsted acid to a neutral molecule and an analogous mechanism (Scheme 2) leads to H/D exchange between the analyte ion and reagent molecules.

The potential energy profiles of Figure 2 also explain why different reagent systems exhibit different abilities to promote exchange. As shown, if the initial proton transfer reaction

$$R'O^- + DRH \rightarrow R'OD + HR^- \qquad (4)$$

is highly exothermic (upper curve), the intermediate energy barrier may be sufficiently high as to prevent the ion/molecule complex A (Scheme 1) reaching the intermediate B in which exchange of the hydrogen for deuterium occurs. In general, H/D exchange between the sample ion and reagent gas will be more efficient when the initial chemical ionization reaction is only slightly

TABLE 1
H/D Isotope Exchange with *m*-Xylene

Reagent Gas	Relative intensity				
	M+2	M+3	M+4	M+5	M+6
D$_2$O[a]	100	53	23	13	7
CH$_3$OD[b]	28	40	57	100	39
C$_2$H$_5$OD[a]	7	7	14	76	100

[a] Data from Hunt and Sethi.[14]
[b] Data from Hawthorne and Miller.[15]

exothermic. Thus, Hunt and Sethi[14] found, in negative ion studies, that D$_2$O exchanged more efficiently with the [M–H]$^-$ ion of toluene than did ND$_3$, while, with the [M – H]$^-$ ion from acetophenone, C$_2$H$_5$OD exchanged more readily than did D$_2$O. These results are in line with the relative gas phase acidities (in kcal mol^{-1}, from Tables 22 and 23, Chapter 2); NH$_3$ (403.7), H$_2$O (390.8), C$_2$H$_5$OH (377.4), C$_6$H$_5$CH$_3$ (380.7), and C$_6$H$_5$COCH$_3$ (361.4). In Brønsted acid CI studies, the data of Table 1 show that C$_2$H$_5$OD (PA = 188.3 kcal mol^{-1}) exchanges more readily with *m*-xylene (PA = 195 kcal mol^{-1}) than does CH$_3$OD (PA = 181.9 kcal mol^{-1}) which in turn exchanges more readily than does D$_2$O (PA = 166.5 kcal mol^{-1}). However, it should be noted that all three reagent gases exchanged equally efficiently with *o*-xylene and *p*-xylene; clearly, structural effects in the substrate molecules also play a role in determining the magnitude of the central barrier in potential energy profiles such as that of Figure 2.

The results to date show that it is possible to establish the number of active hydrogens in a sample molecule without any difficulty by isotope exchange reactions. In favorable cases it may also be possible to establish the number of enolizable hydrogens, the number of benzylic hydrogens or the number of aromatic hydrogens by isotopic exchange reactions, although this approach has seen little application.

III. STEREOCHEMICAL EFFECTS IN CIMS

Electron ionization mass spectrometry has been applied with considerable success to stereochemical problems.[21-24] In general, these studies have shown that stereochemical effects on mass spectra usually are more pronounced at low ion internal energies. For example, Figure 3, Chapter 4, shows that the fragmentation of the molecular ions of the stereoisomeric 4-methylcyclohexanols show much stronger stereochemical effects when formed by low energy charge exchange than when formed by 70 eV electron ionization. Since, in principle at least, chemical ionization can be made a soft ionization technique producing ions of low internal energy, it is to be expected that CIMS should be useful in probing the stereochemistry of the substrate molecules.

Studies of proton transfer to α,ω-difunctional alkanes, such as dimethoxyalkanes,[25] diols[26] and diamines,[27,28] have shown that the proton is particularly strongly bonded to these molecules as the result of intramolecular hydrogen bonding; this internal solvation results in an enhanced PA of the difunctional molecule compared to the corresponding monofunctional molecule. (See, for example, the PAs of diamines and monoamines given in Table 20, Chapter 2). A considerable number of Brønsted acid CI studies have used the presence or absence of internal solvation as a probe of the stereochemistry of the substrate molecule. The occurrence of internal solvation may be signalled by an enhanced stability of the MH$^+$ ion or by an enhanced proportion of simple proton transfer vs. alternative ionization channels (such as clustering) or by both effects.

In early work, Longevialle et al.[29] studied the i-C$_4$H$_{10}$ CI mass spectra of a series of 1,2- and 1,3-aminoalcohols. They observed only MH$^+$ when the configuration of the aminoalcohol permitted intramolecular hydrogen bonding. When such hydrogen bonding was not permitted by the molecular stereochemistry, appreciable fragmentation of MH$^+$ by elimination of H$_2$O was observed. The results obtained correlated well with the evidence for hydrogen bonding in the neutral molecules as deduced from the IR spectra. In an extension of this work Longevialle et al.[30] examined the i-C$_4$H$_{10}$ CI mass spectra of a series of conformationally stable β-aminoalcohols and recorded the % elimination of H$_2$O from MH$^+$ as a function of the dihedral angle between the two functional groups. For angles <90°, no loss of H$_2$O was observed; as the angle increased beyond 90°, the loss of H$_2$O increased reaching at 180°, ~50% for the aminoalcohols and ~60% for the N,N-dimethylaminoalcohols. These values were taken as the respective percentages of protonation at the hydroxyl function. In subsequent work[31] they used the % H$_2$O loss from MH$^+$ to evaluate the conformer populations of conformationally mobile β-aminoalcohols (and N,N-dimethyl derivatives) with results in good agreement with those obtained from IR and NMR studies. They assumed that in conformations where internal solvation was possible, no loss of H$_2$O from MH$^+$ would occur, while in conformations where internal solvation was not possible 50% (aminoalcohols) or 60% (N,N-dimethylaminoalcohols) of the MH$^+$ ions would eliminate H$_2$O. They further assumed that loss of H$_2$O from the hydroxyl-protonated form in the nonsolvating conformation was rapid with respect to conformational change.

There have been many studies which have clearly established that cycloalkane diols and related compounds show substantial differences in their Brønsted acid CI mass spectra dependent on stereochemistry. In general, these differences tend to be more pronounced in i-C$_4$H$_{10}$ CI than in CH$_4$ CI since the former is a gentler ionization method. Winkler and McLafferty[32] first showed that the Brønsted acid CI mass spectra of cycloalkane diols showed substantial stereochemical effects. For the *cis* 1,3- and 1,4-cyclohexanediol isomers, formation of an intramolecular hydrogen bond resulted in much higher MH$^+$ abundances than were observed for the *trans* isomers. As an example, Table 2 records partial i-C$_4$H$_{10}$ CI mass spectra of the 1,4-cyclohexanediols.[33] (The spectra disagree quantitatively with those reported by Winkler and McLafferty although the same conclusions result). The stability of the proton bridge in *cis*-

TABLE 2
Partial i-C_4H_{10} Chemical Ionization Mass Spectra of
1,4 Cyclohexanediols

	Relative intensity	
	cis	trans
M.$C_4H_9^+$	12	42
MH^+	100	58
$[MH-H_2O]^+$	12	100
$[MH-2H_2O]^+$	19	39

1,3-cyclohexanediol is decreased by a sterically interfering 5-methyl group and is enhanced by a sterically interfering 5-hydroxyl group.[32] The Brønsted acid CI mass spectra of *cis* and *trans*-1,2-cyclohexanediol are essentially identical;[32,33] apparently the flexibility of the cyclohexane ring is sufficient to allow internal hydrogen bonding in both isomers. In the more rigid 1,2-cyclopentanediols, the *cis* isomer shows a much more abundant MH^+ than the *trans* isomer.[32] When the conformational flexibility of the 1,2-cyclohexanediols is reduced by forming the trimethylsilyl derivatives, significant differences are observed[34] in the i-C_4H_{10} CI mass spectra with the $[MH - (CH_3)_3SiOH]^+$ ion being much more abundant for the *trans* isomer than for the *cis* isomer. Similarly, the i-C_4H_{10} CI mass spectra of the trimethylsilyl ethers of stereoisomeric 3,4-dimethyl-1,2-cyclopentanediols show substantial differences. The *cis* ethers show abundant MH^+, $[MH - (CH_3)_3SiOH]^+$ and $[MH - 2(CH_3)_3SiOH]^+$ ions while the *trans* ethers show no MH^+ ion signal, the major ion (>95% of ionization) being $[MH - 2(CH_3)_3SiOH]^{+}$.[34] Minor differences in the *cis* diol ether spectra could be correlated with the orientation of the methyl groups. The i-C_4H_{10} CI mass spectra of the trimethylsilyl ethers of the 2,3-dimethyl-1,4-cyclopentanediols show abundant MH^+ ion signals for the *cis* ethers and no MH^+ ion signals for the *trans* ethers.[35]

The i-C_4H_{10} CI mass spectra[36] of the methyl ethers of the 2-methyl-1,3-cyclopentanediols and 2-methyl-1,3-cyclohexanediols (Structures A to F) are

STRUCTURES A-F

TABLE 3

i-C_4H_{10} Chemical Ionization Mass Spectra of Cyclopentanediol and Cyclohexanediol Methyl Ethers

Compound	MH$^+$	[M – H]$^+$	[MH–CH$_3$OH]$^+$	[MH–2CH$_3$OH]$^+$	[MH–46]$^+$
A	0	0.8	94.8	3.0	1.5
B	74.9	0.1	17.5	6.8	0.8
C	59.8	0.3	33.3	5.4	1.2
D	0.7	1.2	38.2	34.1	25.8
E	68.8	0.1	19.8	8.6	2.8
F	51.5	0.3	19.6	16.5	12.2

Note: Intensities are % total ionization.

presented in Table 3. The isomers with the ether functions *trans* to each other show essentially no MH$^+$ ion signal in contrast to the abundant ion signals when the ether groups are *cis* to each other. When the C_2 methyl group also is *cis*, the proton bridge is destabilized. By contrast, when the methyl group is replaced by a t-butyl group[37] the all *cis* isomer gives a more abundant MH$^+$ ion than when the t-butyl group is *trans*, reflecting the instability of the molecule with the t-butyl group in the axial position.

The MH$^+$ ion is abundant in the i-C_4H_{10} CI mass spectrum of 2,5-*diendo*-protadamantanediol due to internal solvation but is of very low abundance in the *diexo* and the two *exo-endo* epimers.[38] Minor differences in fragment ion abundances for the last three epimers could be correlated with structure. In the same vein, the MH$^+$ ion is more abundant in the i-C_4H_{10} CI mass spectra of the diaxial stereoisomers of 2,4-adamantanediol diacetates and dimethyl ethers than in the axial-equatorial and diequatorial stereoisomers.[39] In the i-C_4H_{10} CI mass spectra of [n.3.3]propellane diols, only the *anti,anti* isomer shows an abundant MH$^+$ ion; the *syn,syn* and *syn,anti* isomers may be differentiated by the relative abundances of the [M + C_4H_9 – 2H$_2$O]$^+$ ion signals.[40,41] In the i-butane CI of *syn,anti*-[4.3.3]propellane 8,11 diol it was shown by [18]O labelling that the *syn*-hydroxyl group is preferentially lost in forming the [M – OH]$^+$ ions; it was concluded that this ion originated by a reversible interaction with t-$C_4H_9^+$ ions followed by loss of *tert*-butyl alcohol rather than loss of H$_2$O from the MH$^+$ ions.[40] The stereospecificity for OH loss from M was much reduced in the C_3H_8 CI mass spectrum. From collisional activation experiments, it was concluded[42] that the effect of the *anti* hydroxyl group in promoting the loss of the *syn* hydroxyl was not through an S_Ni mechanism but was expressed either in a weak transition state interaction or by a β-elimination via hydrogen transfer with concomitant double bond formation.

The protonated alcohol function also can be stabilized by interaction with a double bond. Thus, the i-butane CI mass spectra of the *syn* isomers of 7-hydroxynorborn-2-ene and 2-hydroxynorborn-5-ene show a much more abundant MH$^+$ ion signal than is observed for the *anti* isomers.[43,44]

TABLE 4
Chemical Ionization Mass Spectra of 1,2-Cyclopentanediol and 1,2-Cyclohexanediol Acetates

	G		H		I		J	
	CH_4	i-C_4H_{10}	CH_4	i-C_4H_{10}	CH_4	i-C_4H_{10}	CH_4	i-CH_4H_{10}
MH$^+$	1.93	2.1	1.14	5.2	2.13	3.8	1.4	46.5
[MH–HOAc]+	75.6	56.9	21.1	36.2	59.5	52.6	30.2	27.5
[MH–2HOAc]+	6.6	0.4	16.9	4.5	15.3	1.0	14.8	2.3
[MH–HOAc–C_2H_2O]$^+$	4.3	0.2	38.3	3.7	7.7	–	42.4	9.3

Note: Intensities as % of total ionization.

The Brønsted acid CI of diacetate derivatives of cyclic diols have been studied extensively. The CH_4 and i-C_4H_{10} CI mass spectra[45] of the 1,2-cyclopentanediol and 1,2-cyclohexanediol diacetates (Structures G to J) are summarized in Table 4. The i-C_4H_{10} CI mass spectra are characterized by a more abundant MH$^+$ ion signal for the *cis* isomer in each case. The CH_4 CI mass spectra are characterized by a more abundant [MH–HOAc]$^+$ fragment ion and less fragmentation of this species by further loss of HOAc or C_2H_2O for

STRUCTURES G-J

the *trans* isomer. Similar results have been reported for the Brønsted acid CI of the stereoisomeric diacetates of 1,3-cyclopentanediol and 1,3-·cyclohexanediol,[45] 3,4-dimethyl-1,2-cyclopentanediol,[34] 2-methyl-1,3-cyclopentanediol and cyclohexanediol,[36] and 2-*tert*-butyl-1,3-cyclopentanediol and cyclohexanediol.[37] The lower MH$^+$ ion abundances and the unusual stability of the [MH–HOAc]$^+$ species for the *trans* epimers have been attributed to anchimeric assistance (Reaction 5) leading to dioxolanylium and dioxanylium ions.

$$+ \quad HOAc \tag{5}$$

TABLE 5
Partial NH$_3$ Chemical Ionization Mass Spectra of Cycloalkanediols

	[M+NH$_4$]$^+$	MH$^+$
cis-1,3-cyclohexanediol	29	51
trans-1,3-cyclohexanediol	54	2.2
cis-1,4-cyclohexanediol	15	64
trans-1,4-cyclohexanediol	19	0.5
cis-1,2-cyclopentanediol	53	0.4
trans-1,2-cyclopentanediol	56	0.4

Note: Intensities as % total ionization

In related studies it has been shown[46] that *cis* 3-methoxy and 3-trimethylsiloxy cyclopentyl and cyclohexyl acetic acid esters show abundant MH$^+$ ion signals in their i-butane CI mass spectra while the *trans* isomers do not. Only very small differences were observed in the CH$_4$ CI mass spectra. Similarly, *cis*-4-methoxycyclohexane carboxylic acid ethyl ester shows a much more intense MH$^+$ ion signal in the i-C$_4$H$_{10}$ CI mass spectrum than does the *trans* isomer.[47] In the same study, the i-C$_4$H$_{10}$ CI mass spectra of the stereoisomeric 2,6-dimethyl-4-hydroxytetrahydropyranes, 2,4,6-trimethyl-tetrahydropyranes and 2,7-dimethoxy-*cis*-decalins were determined; the MH$^+$ ion was found to be particularly abundant when a proton bridge between the two oxygen functions was permitted by the stereochemistry of the molecule.

When the stereochemistry of the molecule permits intramolecular hydrogen bonding, the PA is increased relative to the epimer where such bonding cannot occur. Thus, with a suitable choice of protonating agent, it should be possible to protonate selectively the epimer of higher PA. The NH$_4$$^+$ ion is a weak protonating agent (PA[NH$_3$] = 204 kcal mol^{-1}) and is known to react by addition when protonation is endothermic. Winkler and Stahl[48] have shown that the NH$_3$ CI mass spectra of cyclic diols reflect their stereochemistry quite strongly. Their results for the epimeric 1,3- and 1,4-cyclohexanediols, and 1,2-cyclopentanediols are presented in Table 5. For the *cis* cyclohexanediols abundant MH$^+$ ion signals are observed; however, for the *trans* isomers, practically no MH$^+$ ions are observed, the NH$_4^+$ ion reacting much more extensively to form [M + NH$_4$]$^+$. The two 1,2-cyclopentanediols show predominant formation of [M + NH$_4$]$^+$. This has been attributed[48] to chelation of the ammonium ion (Structure K) by the 1,2-diol, which is possible for both epimers. Winkler and Stahl noted that in the reaction of C$_4$H$_9^+$ with the 1,2-cyclopentanediols, where such chelation is not possible, the MH$^+$/[M + C$_4$H$_9$]$^+$ ratio was much larger for the *cis* diol than for the *trans* diol. Chen and co-workers have observed that, using trimethyl borate[49] or formaldehyde dimethylacetal[50] as reagent gases, adduct ions are more abundant with *cis* 1,2-cycloalkanediols than with the *trans* epimers.

Functional group interaction as in Structure L, should stabilize the

TABLE 6
F⁻ AND Cl⁻ Chemical Ionization Mass Spectra of Cycloalkanedios

	F⁻ chemical ionization		Cl⁻ chemical ionization	
	[M + F]⁻	[M – H]⁻	[M + Cl]⁻	[M – H]⁻
cis-1,3-cyclohexanediol	10	90	95	4.6
trans-1,3-cyclohexanediol	39	36	99	1.1
cis-1,4-cyclohexanediol	8	92	88	12
trans-1,4-cyclohexanediol	45	44	98	1.7
cis-1,2-cyclopentanediol	12	88	95	4.7
ans-1,.2-cyclopentanediol	42	39	99	1.1

Note: Intensities are % total ionization.

K   **L**

STRUCTURES K AND L

[M – H]⁻ ion of diols and lead to an increased acidity of *cis* diols relative to the *trans* epimers; this difference in acidities should be reflected in the CI mass spectra obtained using weak Brønsted base reagents. Table 6 summarizes the results obtained by Winkler and Stahl[48] for the F⁻ and Cl⁻ CI of a number of epimeric cyclic diols. The Cl⁻ ion reacts predominantly by attachment; however, the [M – H]⁻ ion is more intense for the *cis* diols. This difference is much more pronounced in the F⁻ CI mass spectra where the abundance of [M – H]⁻ for the *cis* diols relative to the *trans* diols reflects the increased acidity of the *cis* epimers. The stabilization of [M – H]⁻ by hydrogen bonding in the *cis* diols also is reflected in the smaller extent of fragmentation of [M – H]⁻ (by elimination of H_2 or H_2O) for the *cis* epimers relative to the *trans* epimers in the OH⁻ CI mass spectra of the diols of Table 6.[51]

In contrast to the compounds discussed above, where interaction of the two functional groups stabilizes the MH⁺ ion, interaction of the two carboxylate groups in dicarboxylic acid derivatives leads a more facile fragmentation of MH⁺.[52,53] As a result, dicarboxylic acid derivatives which have a *cis* configuration about a double bond show more extensive fragmentation in Brønsted acid CI than those which have a *trans* configuration. Results obtained in the CH_4 CI of maleic and fumaric acids and derivatives are summarized in Table 7, where it is seen that the effect is quite dramatic. The thermochemically favored site of protonation in carboxylic compounds is the carbonyl oxygen.[54] For monocarboxylic compounds or *trans* dicarboxylic compounds, elimination of ROH requires (Reaction 6) a symmetry-forbidden 1,3-H migration, a reaction with a high energy barrier.[55] By contrast, interaction of two carboxylate functions provides (Reaction 7) an alternative mode of hydrogen migration

TABLE 7
Partial CH$_4$ Chemical Ionization Mass Spectra of
Dicarboxylic Acid Derivatives

	MH$^+$	[MH–ROH]$^+$
Maleic acid	10	100
Fumaric acid	100	14
Citraconic acid	16	100
Mesaconic acid	100	24
Ethyl maleate	13	100
Ethyl fumarate	100	10[a]
Phenyl maleate	8	100
Phenyl fumarate	100	100

Note: Intensities as % of base peak.
[a] [MH–C$_2$H$_4$]$^+$ = 25

which is not symmetry-forbidden. In addition, the elimination of ROH is favored by formation of the stable cationated anhydride structure.

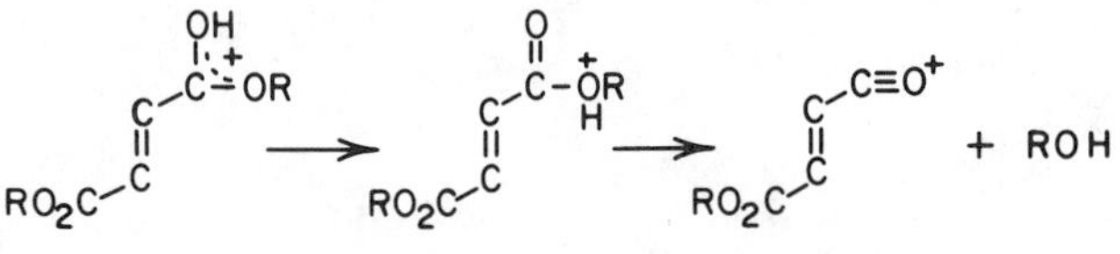

$$(6)$$

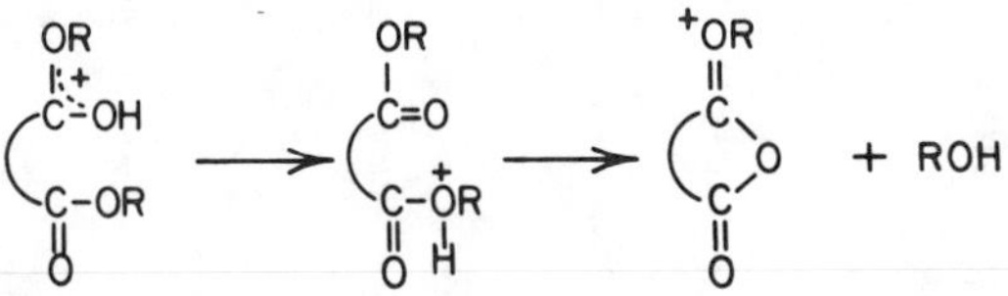

$$(7)$$

The conclusion that, as Reaction 3 shows, the alkyl group eliminated as ROH originates from the carboxylate group which is not protonated in the ionization process has been exploited by Weisz et al.[56] in elegant experiments to probe the site of protonation in isomeric ethyl methyl esters of 2-t-butylmaleic and 2-t-butylsuccinic acids (Structures M and N). As the data in Table 8 show,

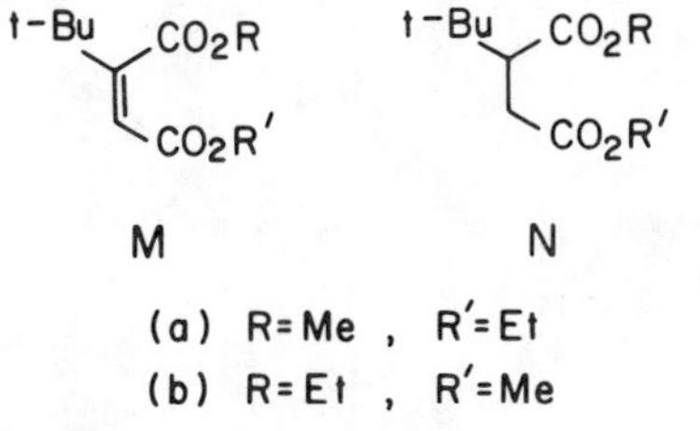

M   N

(a) R=Me , R$'$=Et
(b) R=Et , R$'$=Me

STRUCTURES M and N

TABLE 8
Partial Chemical Ionization Mass Spectra for Ethyl Methyl Maleates and Succinates

Compound	i-C$_4$H$_{10}$ Chemical ionization		NH$_3$ Chemical ionization		CH$_4$ Chemical ionization	
	[MH–MeOH]$^+$	[MH–EtOH]$^+$	[MH–MeOH]$^+$	[MH–EtOH]$^+$	MH–MeOH]$^+$	[MH–EtOH]$^+$
M (a)	100	9	100	5	100	35
M (b)	5	100	<1	100	15	100
N (a)	100	6	49	1	100	22
N (b)	3	100	4	43	14	100

Note: Ion signals as relative intensities. Data from Weisz et al.[56]

the alcohol eliminated from MH^+ contains, with a high specificity, the alkoxy of the carboxylate group adjacent to the t-butyl substituent. Clearly, steric hindrance leads to preferential protonation at the carbonyl oxygen remote from the t-butyl group; on fragmentation, this proton is transferred to the alkoxy group adjacent to the t-butyl group leading to the highly specific elimination of alcohol. As Weisz et al. have pointed out, the PAs of the two ester groups should be similar (possible differences arising from the size of the methyl and ethyl groups are levelled out by comparison of the two isomers); consequently, thermodynamic control of the protonation process should lead to two MH^+ ions and a comparable extent of elimination of the two alcohols. Thus, the results provide clear and direct evidence of kinetic control over the protonation of these compounds.

Stereochemical effects, whose origin often is less clear, frequently are observed in steroidal systems. In early studies, Michnowicz and Munson[57] observed that 5α-androstane-3-one-17β-ol and 5β-androstane-3-one-17β-ol were distinguishable from their H_2 CI mass spectra with the former showing the relative intensities MH^+: $[MH - H_2O]^+$: $[MH - 2H_2O]^+ = 4:2.5:1$ compared to 1:3:3 for the latter epimer. In related work,[58] similar differences were observed in the H_2 CI mass spectra of the epimeric pairs: 5α-androstane-3,17-dione and 5β-androstane-3,17-dione; 5α-cholestane-3-one and 5β-cholestane-3-one. For each pair, the β-epimer showed much more extensive fragmentation by loss of H_2O from MH^+. The results of both studies can be rationalized by noting that the *cis* form of the A/B ring junction in the 5β epimers allows a relatively close interaction between the 3-keto function and the hydrogens on ring B. In more recent studies,[59] it has been observed that, in the CH_4 and NH_3 CI of 5ξ-androstane-3ξ-17ξ-diols, the 17α epimers gave a much more abundant $[MH - 2H_2O]^+$ ion signal than the 17β epimers. In the same study, it was reported that androstane-3-one-17ξ-ols did not show significant differences related to the configuration of the 17-site in their ammonia or methane chemical ionization mass spectra. By contrast, in the OH^- CI mass spectra the 17α-epimers showed a much more intense $[M - H-H_2O]^-$ ion signal than the 17β-epimers, which showed primarily $[M - H]^-$ in their spectra. In the OH^- CI mass spectra of 17ξ-R-$5\alpha,14\beta$-androstane-$14,17\xi$-diols (R = CH_3, C_2H_5, C_2H_3, and C_2H) only the *trans* diol epimers showed loss of RH from the $[M - H]^-$ ion.[60]

Stereochemical effects also are pronounced in the fragmentation of protonated species by retro-Diels-Alder reactions. Zitrin et al.[61] have shown that protonated Structure O (several values of n and m) underwent the retro-Diels-Alder Reaction 8 only for the *cis* isomer.

O

STRUCTURE O

$$(8)$$

In the same vein, Wolfshutz et al.[62] have observed that the retro-Diels-Alder fragmentation reactions of protonated or ethylated Structure P (several $R_1, R_2,$ and R_3), Q, and R are highly specific to the *cis* ring-fused isomers. The fragmentation can be written as Reaction 9 for Structure P with similar formulations for Structures Q and R.

STRUCTURES P-R

$$(9)$$

By contrast, elimination of H_2O from the protonated form of Structure P was more prominent in the *trans* ring-fused epimer.

The regio- and stereo-isomeric bicyclo[3,2,0]heptanone-2 derivatives Structures S to V have been extensively investigated using $i\text{-}C_4H_{10}$ CI mass spectrometry for several series and for R equal acetyl and trimethylsilyl.[63-65] A ready distinction between *hh* and *ht* regio-isomers can be made. The MH$^+$ ions of the *hh* isomers show facile cleavage by Reaction 10 while the *ht*

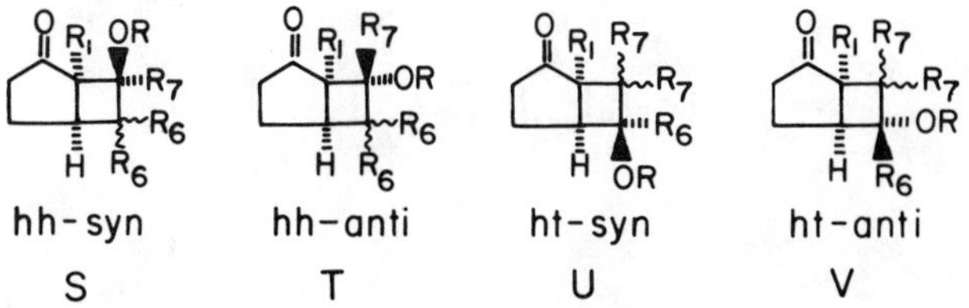

hh-syn   hh-anti   ht-syn   ht-anti

S   T   U   V

STRUCTURE S-V

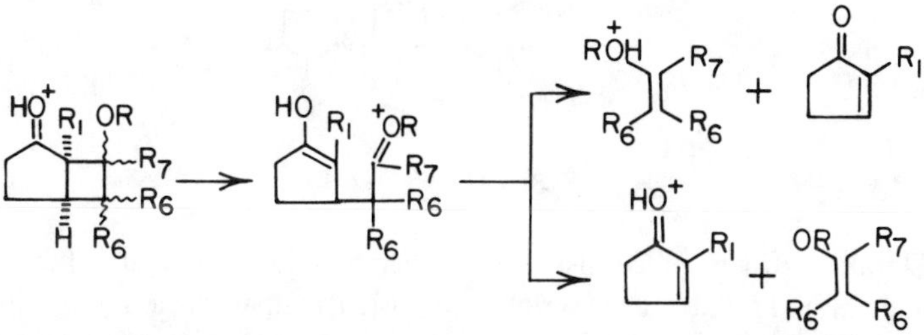

$$(10)$$

regio-isomers fragment to a significant extent by loss of ROH; this latter reaction is more pronounced for the *anti* isomer than the *syn* isomer, which usually shows a more abundant MH⁺ ion signal. When R = Si(CH$_3$)$_3$, the distinction between the *hh-syn* and *hh-anti* isomers was based on the abundance of the ion formed by Reaction 11 which involves migration of R and occurs more readily for the *hh-syn* isomer. When R was an acetyl group, the distinction between *syn* and *anti* isomers was much less certain.

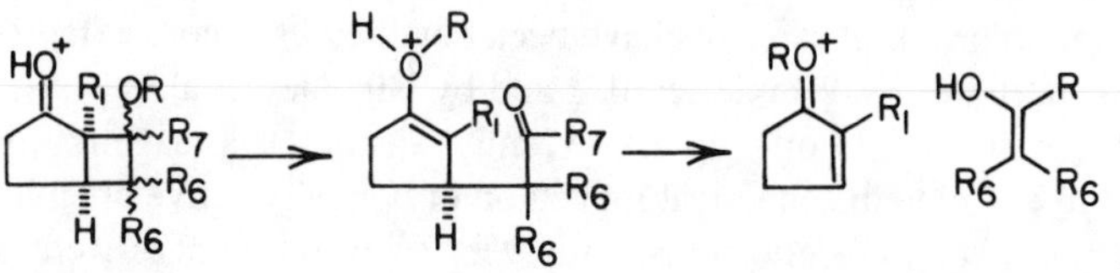

$$(11)$$

The distinction between regio-isomeric *hh* and *ht* species by Brønsted acid CI was achieved earlier[66] for the photodimers W and X. The fission of MH⁺ to give [M/2 + H]⁺ is much more pronounced for the *ht* isomer than for the *hh* isomer since, in the latter case, β-fission forms Y with the charge adjacent to the carbonyl group while β-fission in the *ht* isomer forms the energetically more favorable ion Z (Structures W to Z).

STRUCTURES W-Z

Chirality effects in Brønsted acid chemical ionization also have been reported. In 1977 Fales and Wright[67] reported that, when a 1:1 mixture of diisopropyl-d_0-D-tartrate (M_D) and diisopropyl-d_{14}-L-tartrate (M_L) was introduced into the ion source by direct insertion probe under i-C_4H_{10} CI conditions, the MH$^+$ peaks for the protonated labelled and unlabelled species were approximately equal. However, the ion signals corresponding to the proton-bound dimers $M_D M_D H^+$, $M_D M_L H^+$, and $M_L M_L H^+$ were not in the expected 1:2:1 ratio, that for $M_D M_L H^+$ being only 46% of the expected intensity. When similar experiments were carried out with dimethyl-d_0-D-tartrate, and dimethyl-d_6-D-tartrate the mixed proton-bound dimer was 78% of the expected intensity. Both results indicate a chirality effect in that the mixed dimer containing both the enantiomers is less stable than the dimer containing only one enantiomer. Similar results have been obtained in a more extensive study of dialkyltartrates by Winkler et al.[68] and by Nikolaev et al.[69]

Hua et al.[70] used optically active l-amyl alcohol as a reagent in Brønsted acid CI to examine the chemical ionization of optically active phenylalanines, methionines, and mandelic acids. Only very small differences were observed in the spectra of the enantiomers. In a more recent study, Chen et al.[71] reported that, in the chemical ionization (i-C_4H_{10}) mass spectra of some asymmetric secondary alcohols and α-amino acids, when a pair of enantiomers (such as R- and S-2-phenylbutyric anhydride, R- and S-mandelic acid, R- and S-2-methylbutyric acid, or R- and S-α-phenylethyl amine) were used as a reaction reagent, the abundances of characteristic ions were greater when the sample and reagent had the same configuration. The characteristic ions were proton-bound mixed dimers or fragment ions derived therefrom. Similar results have been reported recently by Martens et al.[72] who found that enantiomers of proline more readily formed a proton-bound mixed dimer with reaction reagents of the same configuration. These startling results must be viewed with some caution, however, since it is not entirely clear, in either of the last two studies, that sufficient precautions were taken to ensure that source concentrations were the same when different enantiomeric pairs were studied.

IV. TANDEM MASS SPECTROMETRY AND CHEMICAL IONIZATION

In the past ten years, there have been many advances in tandem mass spectrometry and two books on the subject have appeared.[73,74] Tandem mass spectrometric experiments, primarily involving collisional studies, have been used to probe the structure of ions formed by chemical ionization, to provide structural information where only molecular mass information is derived from the CI mass spectrum, and to aid in the identification of specific components in complex mixtures. These applications will be reviewed briefly with selected examples. The techniques of tandem mass spectrometry have been outlined in Section VIII, Chapter 3.

Tandem mass spectrometry has been used extensively to probe the site of protonation or cationization in Brønsted acid chemical ionization. Maquestiau et al.[75] found that the ions formed by $C_2H_5^+$ addition to pyrrole gave a high energy collision-induced dissociation mass spectrum which differed from the collision-induced dissociation spectrum of the ions formed by protonation of N-ethylpyrrole. They concluded that, in both cases, the added cation did not bond to the nitrogen. From similarly designed experiments, they concluded that imidazole was protonated or ethylated primarily on the unsubstituted nitrogen, pyridine on the nitrogen, and aniline on the aromatic ring. In contrast to these results, in a similar collision-induced dissociation study, Wood et al.[76] concluded that aniline was both methylated and ethylated predominantly at the nitrogen, although in the ethylation reaction they concluded that ~20% ring attack occurred. Harrison[77] studied, by tandem mass spectrometry, the unimolecular fragmentation reactions occurring after addition of $C_2D_5^+$ to a variety of aromatic molecules containing an unlabelled ethyl group. Unimolecular fragmentation of the adduct with ethylbenzene and ethyltoluene resulted in elimination of C_2H_4 and C_2D_4 in the ratio 2.2:1. This result was interpreted as indicating equivalence of the labelled and unlabelled ethyl groups with an isotope effect favoring elimination of C_2H_4. The adduct of $C_2D_5^+$ and ethyl benzoate also lost C_2H_4 and C_2D_4 in the ratio 2.2:1 indicating equivalence of the two ethyl groups, and thus, ethyl ion addition to the carboxylate function rather than the aromatic ring. The $C_2D_5^+$ adduct with ethylpyridines fragmented primarily by loss of C_2D_4 with only a minor loss of C_2H_4, indicating predominant ethylation at the nitrogen in agreement with Maquestiau. The $C_2D_5^+$ adduct with N-ethylaniline eliminated C_2D_4 and C_2H_4 in the ratio 12.6:1.0, indicating that the two ethyl groups had not become equivalent and suggesting predominant ethylation of the aromatic ring. In agreement with this conclusion, the $C_2D_5^+$ adduct with *p*-ethylaniline eliminated C_2H_4 and C_2D_4 in the ratio 1.3:1.0, indicating significant ethylation of the aromatic ring.

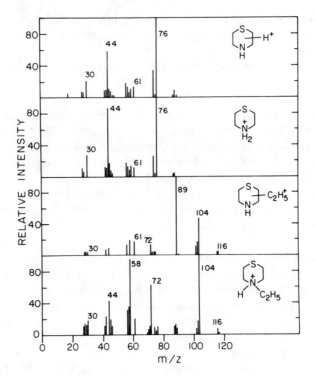

FIGURE 3.   Collision-induced dissociation mass spectra of (from top to bottom) protonated thiomorpholine, m/z 104 from FAB ionization of thiomorpholine hydrochloride, ethylated thiomorpholine, and m/z 132 from FAB ionization of ethylthiomorpholinium iodide. Data from Burinsky and Campana.[78]

Burinsky and Campana[78] have studied the site of protonation, methylation, and ethylation of morpholine, thiomorpholine, and 1,4-thioxane under CI conditions by collisional mass spectrometry. For the nitrogen-containing compounds they compared the collision-induced dissociation spectra of the ions formed by chemical ionization with the collision-induced dissociation of model ions formed by fast atom bombardment (FAB) of suitable quaternary ammonium salts. Figure 3 compares the collision-induced dissociation mass spectra obtained from protonated thiomorpholine with that obtained for the ion of the same m/z obtained by FAB ionization of thiomorpholine hydrochloride. The similarity of the two spectra leaves little doubt that thiomorpholine protonates at the nitrogen under CI conditions. By contrast, as Figure 3 shows, the ion obtained by ethylation of thiomorpholine shows a distinctly different collision-induced dissociation spectrum than the ion of the same m/z obtained by FAB of ethylthiomorpholinium iodide. Clearly, in this case, the ethyl cation has added to the sulfur under CI conditions. Burinsky and Campana concluded that the site of protonation followed the trends in PAs of the heteroatoms (N>S>O) but that the alkyl ion reactivities followed differences in electronegativity or electrophilicity (S>N>O).

Collisional studies in tandem mass spectrometry also are useful in probing the structures of fragment ions formed in chemical ionization reactions. One example will suffice. Maquestiau et al.[79] observed that the $[MH–H_2O]^+$ ion formed in the Brønsted acid CI of *syn*-benzaldoxime gave a collision-induced dissociation mass spectrum similar to that obtained from protonated phenylisocyanide while the $[MH – H_2O]^+$ ion from *anti*-benzaldoxime gave a collision-induced dissociation mass spectrum similar to that obtained for protonated benzonitrile. They concluded that water elimination from MH^+ is concerted with stereospecific migration of the substituent *anti* to the hydroxyl function, Scheme 3. Studies of other oximes were in agreement with this conclusion.

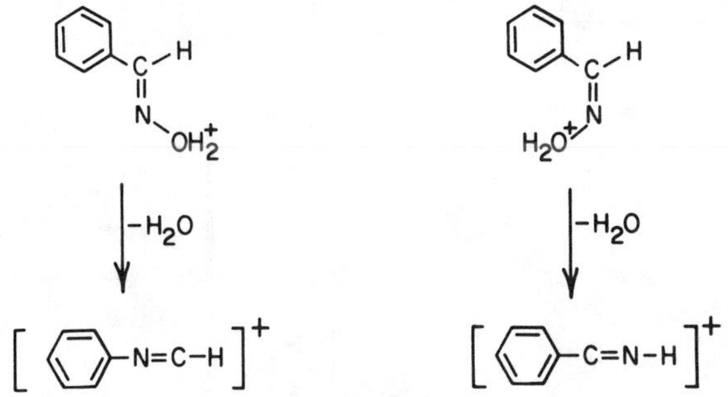

SCHEME 3

Chemical ionization often is a "soft" ionization technique which provides molecular mass information but does not provide structural information; this is particularly true for negative ion chemical ionization. Consequently, collision-induced dissociation studies have seen extensive use in conjunction with chemical ionization in structure elucidation and in distinguishing between isomeric molecules. The potential of such studies was first demonstrated by Soltero-Rigau et al.[80] who showed that the high-energy collision-induced dissociation mass spectra of the MH^+ and $[M + C_2H_5]^+$ ions of a number of barbiturates were distinctive and allowed one to distinguish the isomeric pairs butabarbitol-butethal and pentobarbitol-amobarbitol. Three examples will be presented to illustrate the potential of the method. Figure 4 shows the high-energy negative ion collision-induced dissociation mass spectra of the $[M – H]^-$ ions of three isomeric tripeptides.[81] Although the major fragmentation reaction involves elimination of CO_2 and is uninformative, the lower probability fragmentation reactions involve cleavage at the amide linkages and provide information on the amino acid sequence in these small peptides.

As discussed in Section III, Chapter 5, the location of double bonds in olefinic compounds by mass spectrometric methods is difficult. Figure 5 shows the high energy collision-induced dissociation mass spectrum of the Fe^+ adducts of three isomeric methyl octadecenoates.[82] Although fragment ions are seen at every carbon number, two particularly abundant fragments, labelled A and B, and separated by four carbons are seen in each spectrum. These peaks

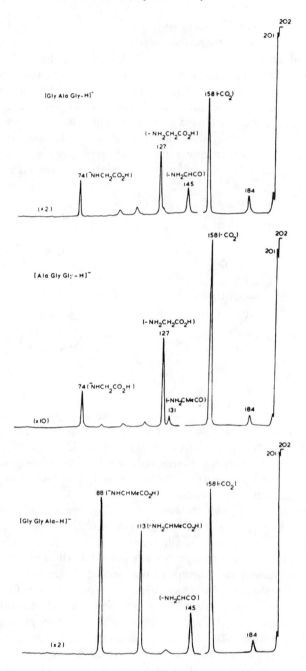

FIGURE 4. Collision-induced dissociation mass spectra of [M – H]⁻ ion from tripeptides GLY-ALA-GLY, ALA-GLY-GLY and GLY-GLY-ALA. (From Eckersley, M., Bowie, J.H., and Hayes, R.N., *Org. Mass Spectrom.*, 24, 597, 1989. With permission.)

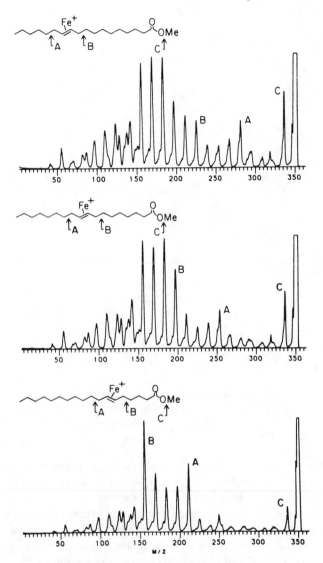

FIGURE 5.   Collision-induced dissociation mass spectra of Fe$^+$ adducts with methyl octadecenoates. (From Peake, D.A. and Gross, M.L., *Anal. Chem.*, 57, 115, 1985. With permission.)

arise by attack of the Fe$^+$ at the allylic C–C bonds with subsequent hydrocarbon elimination. These ions thus serve to indicate the position of the double bond in the hydrocarbon chain.

Even when chemical ionization mass spectra are similar and do not allow isomer distinction, collisional mass spectra may allow such distinction. As the data of Table 6, Chapter 5, show, the O$^-$ CI mass spectra of 3-hexanone and 2-methyl-3-pentanone are very similar. However, as shown in Figure 6, the charge inversion mass spectra of the [M – H]$^-$ ions of the two compounds show

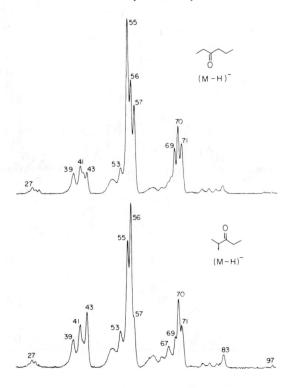

FIGURE 6.   Charge inversion mass spectra of [M – H]⁻ from 3-hexanone and 2-methyl-3-pentanone. (From Marshall, A., Tkaczyk, M., and Harrison, A.G., *J. Am. Soc. Mass Spectrom.*, 2, 292, 1991. With permission.)

substantial differences, particularly in the relative intensities at m/z 55,56, and 57, which allow distinction of the two isomers.[83] In this experiment, the negative ion is chosen by the magnetic sector of a BE instrument (Figure 6, Chapter 3), two electrons are stripped from the ion in the collision process, and the resulting positive ions are mass analyzed by scanning the electric sector.

A particularly useful application of tandem mass spectrometry is in the identification of individual components of complex mixtures without separation of the mixture.[84–87] When a complex mixture is subjected to electron or chemical ionization, a complex mass spectrum results which frequently makes identification of the components impossible. One approach is to separate the components temporally using suitable chromatographic systems coupled to the mass spectrometer, the latter being used to identify each component from its mass spectrum as it elutes from the chromatograph. An alternative approach is to use a multistage mass spectrometer for both separation and identification. In this approach, the complex mixture is ionized, preferably by a soft ionization technique such as CI to simplify the spectrum as far as possible. An ion (MH⁺, [M – H]⁻, etc.) characteristic of the suspected compound is mass selected by the first analysis stage of the tandem mass spectrometer and injected into a collision cell, with the products of the collision-induced reactions being analyzed by the final mass analyzer of the MS/MS system. In effect, the mass separation of the first stage replaces the temporal separation of the chromatograph while the identification from the collision-induced dissociation spectrum replaces identification by the conventional EI or CI mass spectrum.

In early studies, Cooks and co-workers[88-91] used this approach to identify alkaloids in various plant materials and crude extracts. In these studies the material in question was inserted directly into the CI source using a heatable solids probe. As an example, Figure 7 compares the collision-induced dissociation spectrum of an ion of m/z 304 obtained when cocoa leaves were treated in this way with the collision-induced dissociation spectrum obtained from the MH[+] ion of pure cocaine. The close similarity of the two spectra confirms the (expected) presence of cocaine in cocoa leaves.

In a comprehensive study, Hunt and co-workers[92-94] have explored the use of chemical ionization and collision-induced dissociation in a triple quadrupole mass spectrometer for direct analysis of organic contaminants in complex environmental samples. Daughter ion scans, parent ion scans, and neutral loss scans (see Section VIII, Chapter 3) were developed which permitted the direct analysis of ~114 contaminants. For example, the analysis of dialkyl phthalates was based on the observation that all esters higher than the dimethyl ester fragmented as shown in Scheme 4. Thus, in the triple quadrupole instrument,

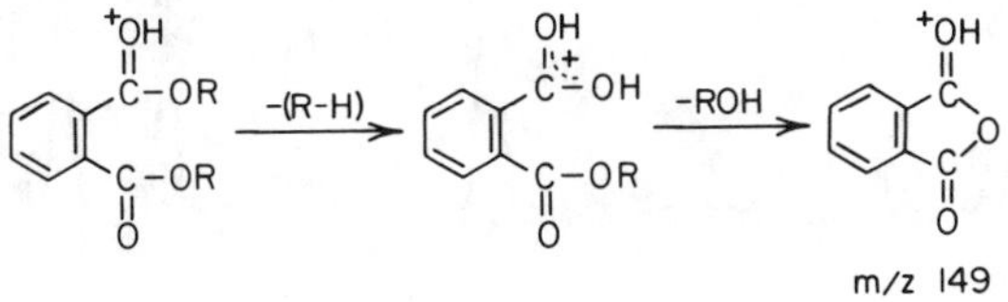

SCHEME 4

the second mass analyzer was set to transmit m/z 149 while the first quadrupole was scanned to admit into the collision cell the precursors which give m/z 149 as a fragment. Typical results for this analysis are shown in Figure 8. Signals identified as m/z 223, 279, 313, and 391 correspond to the MH[+] ions of diethyl, dibutyl, butyl benzyl, and either di-2-ethylhexyl or di-n-octyl phthalate. Fragment ions in the main beam CI (CH_4) mass spectrum at m/z 167, 177, and 205 also are observed in the parent ion scan because they also fragment to produce m/z 149 in the collision cell.

Tandem mass spectrometry also is useful in trace determination of components of complex mixtures even when chromatographic separation is used. The extra stage of mass analysis acts to reduce the chemical noise. Figure 9 illustrates this effect in the analysis of tryptoline in brain extracts as its heptafluorobutyryl derivative using ECCI. The upper trace shows the ion signal recorded for m/z 348 ([M – HF]⁻) for the heptafluorobutyryltryptoline as a function of time during the gas chromatographic run. Clearly, there are many components in the mixture which yield an ion of m/z 348 and the tryptoline derivative, indicated by the arrow, is only a minor component. In the

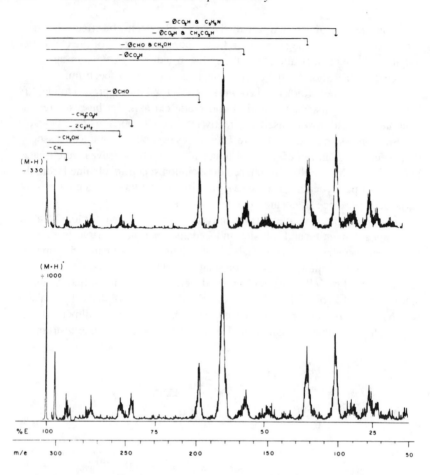

FIGURE 7.   Comparison of collision-induced dissociation spectrum of m/z 304 from cocoa leaves (bottom) with the collision-induced dissociation spectrum of m/z 304 (MH⁺) from authentic cocaine (top). (From Kondrat, R.W. and Cooks, R.G., *Anal. Chem.*, 50, 81A, 1978. With permission.)

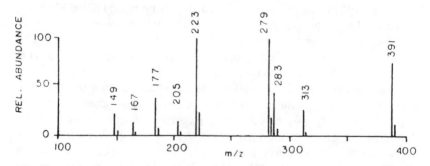

FIGURE 8.   Parent ion scan for formation of m/z 149 from CID of protonated dialkyl phthalates. (From Hunt, D.F., Shabanowitz, J., Harvey, T.M., and Coates, M.L., *J. Chromatogr.*, 271, 93, 1983. With permission.)

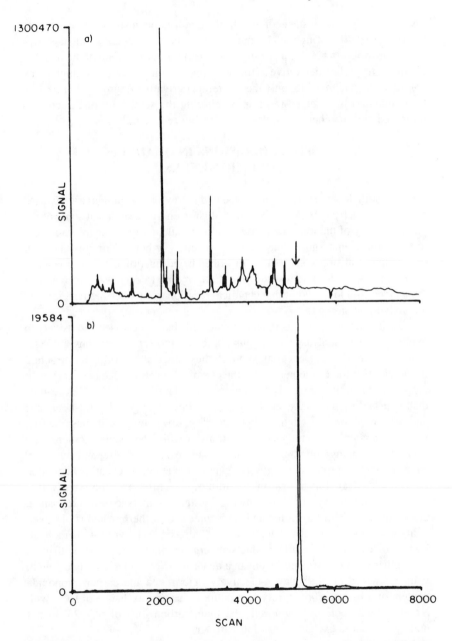

FIGURE 9. Detection of heptafluorobutryl derivative of tryptoline by GC/ECCI. a) Monitoring of m/z 348 during GC run. b) Selected reaction monitoring, m/z 348 → m/z 179 during GC run. (From Johnson, J.V., Yost, R.A., and Faull, K.F., *Anal. Chem.*, 56, 1655, 1984. With permission.)

lower trace, the first stage of the triple quadrupole instrument was set to transmit m/z 348 during the chromatographic run while the second stage was set to transmit m/z 179 ($[C_3F_6COH]^-$) derived by fragmentation of the m/z 348 ion of the tryptoline derivative in the quadrupole collision cell. Although many components give m/z 348, only the tryptoline derivative fragments to m/z 179. Consequently, by selected reaction monitoring the selectivity of detection is improved and the chemical noise considerably reduced.

V. REACTIVE COLLISIONS IN QUADRUPOLE COLLISION CELLS

The collisional processes discussed in the previous section invariably used an inert gas such as He, Ar, or N_2 and collision energies such that some of the kinetic energy of the ion is converted into internal energy of the ion causing it to fragment or undergo a charge change reaction or both. With the advent of quadrupole collision cells as part of either triple quadrupole instruments or hybrid sector/quadrupole instruments, it is possible to decrease the energy of the incident ion to the point that chemical reaction occurs between the ion and a reactive collision gas.

Such reactive collisions in quadrupole cells have been used extensively to explore fundamental aspects of gas-phase ion/molecule reactions.[96–103] Of more interest in the present context are studies aimed at exploring the structure of the incident ion or the neutral by the chemical reactions occurring. Thus, it has been shown that the $C_2H_5O^+$ isomers, CH_3CHOH^+ and $CH_3OCH_2^+$, can be distinguished by the characteristic reactions they undergo with 1,3-butadiene and benzene[104] or with vinyl methyl ether[105] in low energy collisions; the first isomer reacts predominantly by proton transfer while the second isomer reacts primarily by methyl ion transfer. In similar studies, Meyerhoffer and Bursey[106] have shown that protonated *trans*-1,2-cyclopentanediol reacts at low collision energies to transfer a proton to ammonia while protonated *cis*-1,2-cyclopentanediol does not, thus allowing the stereoisomers to be distinguished; proton transfer from the protonated *cis* isomer is endothermic and is observed at higher collision energies. Pachuta et al.[107] also have shown that the molecular ions of a variety of natural products undergo distinctive reactions with ethyl vinyl ether in low energy collisions in a quadrupole cell. In a slightly different vein, Crawford et al.[108] have selectively determined the carbon monoxide content of a hydrocarbon pyrolysis mixture by charge exchange of CO^+ with Kr in a quadrupole collision cell; the hydrocarbon ions of m/z 28 do not undergo a charge exchange reaction with Kr.

An alternative approach is to introduce the analyte into the collision cell and identify it by reaction with mass-selected reactant ions. In this approach, Meyerhoffer and Bursey[109] have shown that $(CH_3)_3Si^+$ reacts selectively with *cis*-1,2-cyclopentanediol at low collision energies to produce $(CH_3)_3SiOH_2^+$; the reaction with the *trans* isomer is endothermic and does not occur until higher collision energies are reached. Hail et al.[110] have interfaced a gas chromatograph to the collision cell of a triple quadrupole instrument and have shown that the aromatic components of a test mixture are selectively ionized as they elute from the gas chromatograph by using $C_6H_6^+$ as a charge exchange reagent ion. All components of the mixture were detected using Ar^+ as a charge exchange reagent ion. In the same study, they showed that selective and distinctive ionization of their test mixture occurred using the proton-bound dimer of acetone as a reagent ion.

There is considerable advantage in being able to mass-select the reactant ion in chemical ionization reactions as the above examples illustrate. As pointed out in Chapter 3, a similar selection can be made with FTMS and quadrupole ion trap mass spectrometers by application of suitable pulse sequences. Berberich et al.[111] recently have published an extensive examination of mass-selection of reactant ions in quadrupole ion trap and triple quadrupole mass spectrometers.

REFERENCES

1. **Biemann, K.**, *Mass Spectrometry: Organic Chemical Applications,* McGraw-Hill, New York, 1962.
2. **Budzikiewicz, H., Djerassi, C., and Williams, D.H.**, *Structure Elucidation of Natural Products by Mass Spectrometry,* Vol. 1, Holden-Day, San Francisco, 1964.
3. **Holmes, J.L.**, Isotopic labelling as a tool for determining fragmentation mechanisms, in Mass Spectrometry, Maccoll, A., Ed., International Review of Science, Physical Chemistry, Series 2, Butterworths, London, 1975
4. **Hunt, D.F., McEwen, C.N., and Upham, R.A.**, Determination of active hydrogen in organic compounds by chemical ionization mass spectrometry, *Anal. Chem.,* 44, 1292, 1972.
5. **Hunt, D.F., McEwen, C.N., and Upham, R.A.**, Chemical ionization mass spectrometry. II. Differentiation of primary, secondary and tertiary amines, *Tetrahedron Lett.,* 4539, 1971.
6. **Blum, W., Schrumpf, E., Liehr, J.G., and Richter, W.**, On-line hydrogen/deuterium exchange in capillary gas chromatography-chemical ionization mass spectrometry (GC-CIMS) as a means of structure analysis in complex mixtures, *Tetrahedron Lett.,* 565, 1976.
7. **Lin, Y.Y. and Smith, L.L.**, Active hydrogens by chemical ionization mass spectrometry, *Biomed. Mass Spectrom.,* 6, 15, 1979.
8. **Buchanan, M.V.**, Mass spectral characterization of nitrogen-containing compounds with ammonia chemical ionization, *Anal. Chem.,* 54, 570, 1982.
9. **Buchanan, M.V.**, Mass spectral characterization of oxygen-containing aromatics with methanol chemical ionization, *Anal. Chem.,* 56, 546, 1984.
10. **Liu, R.H., Low, I.A., Smith, F.P., Piotrowski, E.G., and Hsu, A.–F.**, Chemical ionization mass spectrometric characterization of morphine alkaloids, *Org. Mass Spectrom.,* 20, 511, 1985.
11. **Miller, D.J. and Hawthorne, S.B.**, Identification of heteroaromatic compounds in coal liquefaction and gasification products by deuterated reagent chemical ionization mass spectrometry, *Fuel,* 68, 105, 1989.
12. **Freiser, B.S., Woodin, R.L., and Beauchamp, J.L.**, Sequential deuterium exchange reactions of protonated benzenes with D_2O in the gas phase by ion cyclotron resonance spectroscopy, *J. Am. Chem. Soc.,* 97, 6895, 1975.
13. **Martinson, D.P. and Butrill, S.E.**, Determination of the site of protonation of substituted benzenes in water chemical ionization mass spectrometry, *Org. Mass Spectrom.,* 11, 762, 1976.

14. **Hunt, D.F. and Sethi, S.K.**, Gas-phase ion/molecule isotope exchange reactions: methodology for counting hydrogen atoms in specific structural environments by chemical ionization mass spectrometry, *J. Am. Chem. Soc.*, 102, 6953, 1980.

15. **Hawthorne, S.B. and Miller, D.J.**, Identifying alkylbenzene isomers with chemical ionization-proton exchange mass spectrometry, *Anal. Chem.*, 57, 694, 1985.

16. **Miller, D.J. and Hawthorne, S.B.**, Chemical ionization-proton exchange mass spectrometric identification of alkyl benzenes, tetralins and indans in a coal-derived jet fuel, in *Novel Techniques in Fossil Fuel Mass Spectrometry*, ASTM STP 1019, Ashe, T.R. and Wood, K.V., Eds., Am. Soc. Test. Mat., Philadelphia, 1989.

17. **Stewart, J.H., Shapiro, R.H., DePuy, C.H., and Bierbaum, V.M.**, Hydrogen-deuterium exchange reactions of carbanions with D_2O in the gas phase, *J. Am. Chem. Soc.*, 99, 7650, 1977.

18. **DePuy, C.H., Bierbaum, V.M., King, G.K., and Shapiro, R.H.**, Hydrogen-deuterium exchange reactions of carbanions with deuterated alcohols in the gas phase, *J. Am. Chem. Soc.*, 100, 2921, 1978.

19. **Brauman, J.I.**, Factors influencing thermal ion-molecule rate constants, in *Kinetics of Ion-Molecule Reactions,* Ausloos, P., Ed., Plenum Press, New York, 1979.

20. **Magnera, T.F. and Kebarle, P.**, Exothermic bimolecular ion molecule reactions with negative temperature dependences, in *Ionic Processes in the Gas Phase*, Almoster Ferreira, M.A., Ed., Reidel, Dordrecht, 1984.

21. **Green, M.**, Mass spectrometry and the stereochemistry of organic molecules, in *Topics in Stereochemistry*, Vol. 9, Allinger, N.L. and Eliel, E.L., Eds., John Wiley & Sons, New York, 1976.

22. **Mandelbaum, A.**, Application of mass spectrometry to stereochemical problems, in *Stereochemistry, Fundamentals and Applications*, Vol. 1, Kagan, H.B., Ed., Georg Thieme Verlag, Stuttgart, 1977.

23. **Green, M.**, Mass spectrometry — a sensitive probe of molecular geometry, *Pure Appl. Chem.,* 50, 185, 1978.

24. **Mandelbaum, A.**, Stereochemical effects in mass spectrometry, *Mass Spectrom. Rev.*, 2, 223, 1983.

25. **Morton, T.L. and Beauchamp, J.L.**, Chemical consequences of strong hydrogen bonding in the reactions of organic ions in the gas phase. Interactions of remote functional groups, *J. Am. Chem. Soc.*, 94, 3671, 1972.

26. **Dzidic, I. and McCloskey, J.A.**, Influence of remote functional groups in the chemical ionization of long chain compounds, *J. Am. Chem. Soc.,* 93, 4955, 1971.

27. **Aue, D.H., Webb, H.M., and Bowers, M.T.**, Quantitative evaluation of intramolecular strong hydrogen bonding in the gas phase, *J. Am. Chem. Soc.*, 95, 2699, 1973.

28. **Yamdagni, R. and Kebarle, P.**, Gas phase basicities of amines. Hydrogen bonding in proton-bound amine dimers and proton-induced cyclization of α,ω-diamines, *J. Am. Chem. Soc.*, 95, 3504, 1973.

29. **Longevialle, P., Milne, G.W.A., and Fales, H.M.**, Chemical ionization mass spectrometry of complex molecules. XI. Stereochemical and conformational effects in the isobutane chemical ionization mass spectra of some steroidal amino-alcohols, *J. Am. Chem. Soc.*, 95, 6666, 1973.

30. **Longevialle, P., Girard, J.–P., Rossi, J.–C., and Tichy, M.**, Influence of the interfunctional distance on the dehydration of amino alcohols in isobutane chemical ionization mass spectrometry, *Org. Mass Spectrom.*, 14, 414, 1979.

31. **Longevialle, P., Girard, J.–P., Rossi, J.–C., and Tichy, M.**, Isobutane chemical ionization mass spectrometry of β-amino-alcohols. Evaluation of populations of conformers at equilibrium in the gas phase, *Org. Mass Spectrom.*, 15, 268, 1980.

32. **Winkler, J. and McLafferty, F.W.**, Stereochemical effects in the chemical ionization mass spectra of cyclic diols, *Tetrahedron*, 30, 2977, 1974.

33. **Munson, B.**, Chemical ionization mass spectrometry — analytical applications of ion-molecule reactions, in *Interactions Between Ions and Molecules*, Ausloos, P., Ed., Plenum Press, New York, 1975.

34. **Van de Sande, C.C., Van Gaever, F., Hanselaer, R., and Vandewalle, M.**, Correlation between the stereochemistry and the chemical ionization spectra of epimeric 3,4-dimethyl-1,2-cyclopentanediols, *Z. Naturforsch.*, 32b, 810, 1977.

35. **Claeys, M. and Van Haver, D.**, Conformation effects in the chemical ionization spectra of derivatives of isomeric 2,3-dimethyl-1,4-cyclopentanediols, *Org. Mass Spectrom.*, 12, 531, 1977.

36. **D'Haenens, L., Van de Sande, C.C., and Vangaever, F.**, The isobutane chemical ionization mass spectra of 2-methyl substituted 1,3-cyclopentanediol diacetates and dimethyl ethers, *Org. Mass Spectrom.*, 14, 145, 1979.

37. **D'Haenens, L., Van de Sande, C.C., Schelfaut, M., and Vandewalle, M.**, The chemical ionization mass spectra of 2-tert-butyl-substituted 1,3-cycloalkane diacetates and dimethyl ethers, *Org. Mass Spectrom.*, 22, 330, 1987.

38. **Munson, B., Jelus, B., Hatch, F., Morgan, T.K., and Murray, R.K.**, Stereochemical effects in the mass spectra of 2-hydroxy-, 5-hydroxy-, and 2,5-dihydroxyadamantanes, *Org. Mass Spectrom.*, 15, 161, 1980.

39. **D'Haenens, L., Van de Sande, C.C., and Schelfaut, M.**, The chemical ionization mass spectra of 2,4-adamantanediol diacetates and dimethyl ethers, *Org. Mass Spectrom.*, 17, 265, 1982.

40. **Ashkenazi, P., Blum, W., Domon, B., Gutman, A.–L., Mandelbaum, A., Müller, D., Richter, W.J., and Ginsberg, D.**, Propellanes 91. Fragmentation mechanisms of alcohols under isobutane chemical ionization. Highly stereospecific formation of $[M–OH]^+$ ions from [4.3.3]propellane-8,11-diols, *J. Am. Chem. Soc.*, 109, 7325, 1987.

41. **Klopstock, Y., Ashkenazi, P., Mandelbaum, A., Müller, D., Richter, W.J., and Ginsberg, D.**, Propellanes XCVI. Unambiguous chemical ionization mass spectral assignment of configurations to the diols obtained by reduction of [n.3.3]propellanediones and HPLC purification, *Tetrahedron*, 44, 5893, 1988.

42. **Ashkenazi, P., Domon, B., Gutman, A.–L., Mandelbaum, A., Müller, D., Richter, W.J., and Ginsberg, D.**, Propellanes 100. The nature of the stereoelectronic effect in the highly stereospecific hydroxyl elimination from *syn,anti*-[4.3.3]propellane-8,11-diol upon isobutane chemical ionization, *Israel J. Chem.*, 29, 131, 1989.

43. **Jalonen, J. and Taskinen, J.**, Organic chemistry of gas-phase ions. Part 1. Effect of the protonation site in stereoisomeric norborneols, *J. Chem. Soc. Perkin Trans.* 2, 1833, 1985.

44. **Munson, B., Feng, T.–M., Ward, H.D., and Murray, Jr., R.K.**, Isobutane chemical ionization mass spectra of unsaturated alcohols, *Org. Mass Spectrom.*, 22, 606, 1987.

45. **Respondek, J., Schwarz, H., Van Gaever, F., and Van de Sande, C.C.**, 1,3-Dioxolanylium und 1,3-dioxanylium-derivate via protenkatalysierte S_N1 reacktionen in der gas phase, *Org. Mass Spectrom.*, 13, 618, 1978.

46. **Van Gaever, F., Monstrey, J., and Van de Sande, C.C.**, Chemical ionization mass spectrometry of bifunctional cyclopentanes and cyclohexanes. A correlation between stereochemistry and chemical ionization spectra, *Org. Mass Spectrom.*, 12, 200, 1977.

47. **Van de Sande, C.C., Van Gaever, F., Sandra, P., and Monstrey, J.**, Chemical ionization mass spectrometry: a useful stereochemistry probe, *Z. Naturforsch.*, 32b, 573, 1977.

48. **Winkler, F.J. and Stahl, D.**, Intramolecular ion solvation effects on gas-phase acidities and basicities. A new stereochemical probe in mass spectrometry, *J. Am. Chem. Soc.*, 101, 3685, 1979.

49. **Hua, S., Chen, Y., Jiang, L., and Xue, S.**, Stereochemical effects in mass spectrometry. 2. Chemical ionization mass spectra of cyclic glycols and mono- and di-saccharides using trimethyl borate as reagent gas, *Org. Mass Spectrom.*, 20, 719, 1985.

50. **Tu, Y.–P., Chen, Y.–Z., Chen, S.–N., Wong, M.–L., and Jing, Z.–Z.,** Stereochemical effects in mass spectrometry. Part 8. Chemical ionization mass spectra of cyclic glycols and monosaccharides using formaldehyde dimethyl acetal as reagent gas, *Org. Mass Spectrom.,* 25, 9, 1989.

51. **Winkler, F.J. and Stahl, D.,** Stereochemical effects on anion mass spectra of cyclic diols. Negative chemical ionization, collisional activation, and metastable ion spectra, *J. Am. Chem. Soc.,* 100, 6779, 1978.

52. **Weinkam, R.J. and Gal, J.,** Effects of bifunctional interactions in the chemical ionization mass spectrometry of carboxylic acids and methyl esters, *Org. Mass Spectrom.,* 11, 188, 1976.

53. **Harrison, A.G. and Kallury, R.K.M.R.,** Stereochemical applications of mass spectrometry. II. Chemical ionization mass spectra of isomeric dicarboxylic acids and derivatives, *Org. Mass Spectrom.,* 15, 277, 1980.

54. **Benoit, F.M. and Harrison, A.G.,** Predictive value of proton affinity-ionization energy correlations involving oxygen-containing molecules, *J. Am. Chem. Soc.,* 99, 3980, 1977.

55. **Middlemiss, N.E. and Harrison, A.G.,** Structure and fragmentation of gaseous protonated acids, *Can. J. Chem.,* 57, 2827, 1979.

56. **Weisz, A., Cojocaru, M., and Mandelbaum, A.,** Site specific gas-phase protonation of 2-t-butylmaleates and 2-t-butylsuccinates upon chemical ionization: stereochemical effects and kinetic control, *J. Chem. Soc. Chem. Commun.,* 331, 1989.

57. **Michnowicz, J. and Munson, B.,** Studies in chemical ionization mass spectrometry: 17-hydroxy steroids, *Org. Mass Spectrom.,* 6, 765, 1972.

58. **Michnowicz, J. and Munson, B.,** Studies in chemical ionization mass spectrometry: steroidal ketones, *Org. Mass Spectrom.,* 8, 49, 1974.

59. **Prome, D., Prome, J.–C., and Stahl, D.,** Distinction between 17-epimeric hydroxy steroids of the 3,17-dioxygenated androstane series by chemical ionization, *Org. Mass Spectrom.,* 20, 525, 1985.

60. **Beloeil, J.C., Bertranne, M., Stahl, D., and Tabet, J.C.,** Stereochemistry of gaseous anions: OH⁻ negative chemical ionization of 17ξ-R-5α,14β-androstane-14,17ξ-diols, *J. Am. Chem. Soc.,* 105, 1355, 1983.

61. **Zitrin, S., Yinon, J., and Mandelbaum, A.,** Stereochemistry of retro-Diels-Alder fragmentation in gas-phase ions formed by chemical ionization, *Tetrahedron,* 34, 1199, 1978.

62. **Wolfshutz, R., Gransee, M., Seedorf, M., and Schwarz, H.,** Stereochemische effecte bei sauerkatalysierten retro-Diels-Alder reaktionen in der gasphase — eine neue analytische anwendung der chemische ionisation, *Z. Anal. Chem.,* 295, 143, 1979.

63. **Termont, D., Van Gaever, F., Dekeukeleire, D., Claeys, M., and Vandewalle, M.,** Differentiation between regio- and stereo-isomers of bicyclo[3,2,0]heptanone-2 derivatives by chemical ionization mass spectrometry, *Tetrahedron,* 33, 2433, 1977.

64. **Claeys, M., Matveeva, H., Devreese, A., Termont, D., and Vandewalle, M.,** Regio and stereochemical effects in the chemical ionization mass spectra of isomeric bicyclo[3,2,0]heptan-2-one trimethylsiloxy derivatives, *Bull. Soc. Chim. Belg.,* 87, 375, 1978.

65. **Claeys, M., Van Audenhove, M., and Vanderwalle, M.,** Evidence for the occurrence of an oxygen to acyl shift in 7-acetoxybicyclo[3,2,0]heptan-2-ones upon chemical ionization, *Bull. Soc. Chim. Belg.,* 88, 799, 1979.

66. **Ziffer, H., Fales, H.M., Milne, G.W.A., and Field, F.H.,** Chemical ionization mass spectrometry of complex molecules. III. The structures of the photodimers of α,β-unsaturated ketones, *J. Am. Chem. Soc.,* 92, 1597, 1970.

67. **Fales, H.M. and Wright, G.J.,** Detection of chirality with the chemical ionization mass spectrometer. "Meso" ions in the gas phase, *J. Am. Chem. Soc.,* 99, 2339, 1977.

68. **Winkler, F.J., Stahl, D., and Maquin, F.,** Chirality effects in the chemical ionization mass spectra of dialkyltartrates, *Tetrahedron Letts.,* 27, 335, 1986.

69. **Nikolaev, E.N., Gaginashvili, G.I., Tal'rose, V.L., and Kostyanovsky, R.G.**, Investigation of asymmetric gas-phase ion/molecule reactions by FT-ICR spectrometry, *Int. J. Mass Spectrom. Ion Processes*, 86, 249, 1988.

70. **Hua, S., Chen, Y., Jiang, L., and Xue, S.**, Stereochemical effects in mass spectrometry. 3. Detection of chirality by chemical ionization mass spectrometry, *Org. Mass Spectrom.*, 21, 7, 1986.

71. **Chen, Y.–Z., Li, H., Yang, H.–J., Hua, S.–M., Li, H.–G., Zhao, F.–Z. and Chen, N.–Z.**, Stereochemical effects in mass spectrometry. 7. Determination of absolute configuration of some organic molecules by reaction mass spectrometry, *Org. Mass Spectrom.*, 23, 821, 1988.

72. **Martens, J., Lübben, S., and Schwarting, W.**, Stereoselective reaction mass spectrometry with cyclic α-amino acids, *Z. Naturforschung*, 46b, 320, 1991.

73. **McLafferty, F.W.**, Ed., *Tandem Mass Spectrometry*, John Wiley & Sons, New York, 1983.

74. **Busch, K.L., Glish, G.L., and McLuckey, S.A.**, *Mass Spectrometry/Mass Spectrometry: Techniques and Applications of Tandem Mass Spectrometry*, VCH Publishers, New York, 1988.

75. **Maquestiau, A., Van Haverbeke, Y., Mispreuve, H., Flammang, R., Harris, J.A., Howe, I., and Beynon, J.H.**, The gas phase structure of some protonated and ethylated amines, *Org. Mass Spectrom.*, 15, 144, 1980.

76. **Wood, K.V., Burinsky, D.J., Cameron, D., and Cooks, R.G.**, Site of gas-phase cation attachment. Protonation, methylation and ethylation of aniline, phenol and thiophenol, *J. Org. Chem.*, 48, 5236, 1983.

77. **Harrison, A.G.**, Site of gas-phase ethyl ion attachment, *Can. J. Chem.*, 64, 1051, 1986.

78. **Burinsky, D.J. and Campana, J.E.**, The site of gas phase cation attachment. The protonation, methylation and ethylation of morpholine, thiomorpholine and 1,4-thioxane, *Org. Mass Spectrom.*, 19, 539, 1984.

79. **Maquestiau, A., Van Haverbeke, Y., Flammang, R., and Meyrant, P.**, Chemical ionization mass spectrometry of some aromatic and aliphatic oximes, *Org. Mass Spectrom.*, 15, 80, 1980.

80. **Soltero–Rigau, E., Kruger, T.L., and Cooks, R.G.**, Identification of barbiturates by chemical ionization and mass analyzed ion kinetic energy spectrometry, *Anal. Chem.*, 49, 435, 1977.

81. **Eckersley, M., Bowie, J.H., and Hayes, R.N.**, Collision-induced dissociations of deprotonated peptides, dipeptides and tripeptides with hydrogen and alkyl alpha groups. An aid to structure determination, *Org. Mass Spectrom.*, 24, 597, 1989.

82. **Peake, D.A. and Gross, M.L.**, Iron (I) chemical ionization and tandem mass spectrometry for locating double bonds, *Anal. Chem.*, 57, 115, 1985.

83. **Marshall, A., Tkaczyk, M., and Harrison, A.G.**, O⁻ chemical ionization of carbonyl compounds, *J. Am. Soc. Mass Spectrom.*, 2, 292, 1991.

84. **McLafferty, F.W. and Bockhoff, F.M.**, Separation/identification for complex mixtures using mass separation and mass spectral characterization, *Anal. Chem.*, 50, 69, 1978.

85. **Kondrat, R.W. and Cooks, R.G.**, Direct analysis of mixtures by mass spectrometry, *Anal. Chem.*, 50, 81A, 1978.

86. **Cooks, R.G.**, Mixture analysis by mass spectrometry, in *Trace Organic Analysis: A New Frontier in Analytical Chemistry*, NBS Spec. Publ. 519, Hertz, H.S. and Chester, S.N., Eds., U.S. Department of Commerce, Washington, 1979.

87. **McLafferty, F.W.**, Tandem mass spectrometry (MS/MS): a promising new technique for specific component determination in complex mixtures, *Acc. Chem. Res.*, 13, 33, 1980.

88. **Kruger, T.L., Cooks, R.G., McLaughlin, J.L., and Ranieri, R.L.**, Identification of alkaloids in crude extracts by mass analyzed ion kinetic energy. spectrometry, *J. Org. Chem.*, 42, 4161, 1977.

89. **Kondrat, R.W., Cooks, R.G., and McLaughlin, J.L.**, Alkaloids in whole plant material: direct analysis by kinetic energy spectrometry, *Science*, 199, 978, 1978.

90. **Kondrat, R.W., McCluskey, G.A., and Cooks, R.G.**, Multiple reaction monitoring in mass spectrometry/mass spectrometry for direct analysis of complex mixtures, *Anal. Chem.*, 50, 2017, 1978.

91. **Youssefi, M., Cooks, R.G., and McLaughlin, J.L.**, Mapping of cocaine and cinnamoylcocaine in whole coca plant tissues by MIKES, *J. Am. Chem. Soc.*, 101, 3400, 1979.

92. **Hunt, D.F., Shabanowitz, J., and Giordani, A.B.**, Collision activated decomposition of negative ions in mixture analysis with a triple quadrupole mass spectrometer, *Anal. Chem.*, 52, 386, 1980.

93. **Hunt, D.F., Shabanowitz, J., Harvey, T.M., and Coates, M.L.**, Analysis of organics in the environment by functional group using a triple quadrupole mass spectrometer, *J. Chromatogr.*, 271, 93, 1983.

94. **Hunt, D.F., Shabanowitz, J., Harvey, T.M., and Coates, M.**, Scheme for the direct analysis of organics in the environment by tandem mass spectrometry, *Anal. Chem.*, 57, 525, 1985.

95. **Johnson, J.V., Yost, R.A., and Faull, K.F.**, Tandem mass spectrometry for the trace determination of tryptolines in crude brain extracts, *Anal. Chem.*, 56, 1655, 1984.

96. **Mitchell, A.L., and Tedder, J.M.**, The study of ion molecule reactions in the gas phase using a triple quadrupole mass spectrometer. Part 2. The reaction of CH_3^+ and CH_4^+ with linear, branched and cyclic alkanes, *J. Chem. Soc. Perkin Trans.* 2, 667, 1984.

97. **Schmit, J.P., Dawson, P.H., and Beaulieu, N.**, Chemical synthesis inside the collision cell of a MS/MS system. 1. Formation of adduct ions between protonated esters and ammonia, *Org. Mass Spectrom.*, 20, 269, 1985.

98. **Schmit, J.–P., Beaudet, S., and Brisson, A.**, Ion/molecule reactions inside the collision cell of a tandem quadrupole mass spectrometric system. 2. Energy dependence of ammonium ion formation, *Org. Mass Spectrom.*, 21, 493, 1986.

99. **Glish, G.L., McLuckey, S.A., McBay, E.H., and Bertram, L.K.**, Design and performance of a hybrid mass spectrometer of QEB geometry, *Int. J. Mass Spectrom. Ion Processes*, 70, 321, 1986.

100. **Mitchell, A.L. and Tedder, J.M.**, Studies of ion molecule reactions using a multiple quadrupole mass spectrometer. Part 3. The reactions of alkenyl radical cations $C_nH_{2n}^+$ with alkenes, *J. Chem. Soc. Perkin Trans.* 2, 1197, 1986.

101. **Stanney, K., Tedder, J.M., and Mitchell, A.L.**, A study of ion molecule reactions in the gas phase. Part 3. The reactions of methyl and fluoromethyl cations with alkenes and fluoroalkenes in the gas phase, *J. Chem. Soc. Perkin Trans.* 2, 1383, 1986.

102. **Robinson, J.N., and Tedder, J.M.**, The reactions of ions $[XC]^+$, $[X_2C]^{+\cdot}$, $[X_3C]^+$, and $[X_4C]^{+\cdot}$ (when X = H or F) with ethene and propene, *Org. Mass Spectrom.*, 22, 154, 1987.

103. **White, E.L., Tabet, J.C., and Bursey, M.M.**, Reaction of ammonia with accelerated benzoyl ions under multiple-collision conditions in a triple quadrupole instrument, *Org. Mass Spectrom.*, 22, 132, 1987.

104. **Jalonen, J.**, Application of reactive collisions for differentiation of isomeric organic ions in the gas phase, *J. Chem. Soc. Chem. Commun.*, 872, 1985.

105. **Harrison, A.G. and Young, A.B.**, Studies in gaseous ion chemistry using a hybrid BEQQ mass spectrometer, *Int. J. Mass Spectrom. Ion Processes*, 94, 321, 1989.

106. **Meyerhoffer, W.J. and Bursey, M.M.**, Differentiation of the isomeric 1,2-cyclopentanediols by ion-molecule reactions in a triple quadrupole mass spectrometer, *Org. Mass Spectrom.*, 24, 169, 1989.

107. **Pachuta, R.R., Kenttämaa, H.I., Cooks, R.G., Zennie, T.M., Ping, C., Ching, C.–J., and Cassady, J.M.**, Analysis of natural products by tandem mass spectrometry employing reactive collisions with ethyl vinyl ether, *Org. Mass Spectrom.*, 23, 10, 1988.

108. **Crawford, R.W., Alcaraz, A., and Reynolds, J.G.**, Real-time, on-line determination of carbon monoxide using charge exchange with krypton in a triple quadrupole mass spectrometer, *Anal. Chem.*, 60, 2439, 1988.

109. **Meyerhoffer, W.J. and Bursey, M.M.**, Reactions of the trimethylsilyl ion with 1,2-cyclopentanediol isomers in the collision region of a triple quadrupole instrument, *Org. Mass Spectrom.*, 24, 246, 1989.

110. **Hail, M.E., Berberich, D.W., and Yost, R.A.**, Gas chromatographic sample introduction into the collision cell of a triple quadrupole mass spectrometer for mass selection of reactant ions for charge exchange and chemical ionization, *Anal. Chem.*, 61, 1874, 1989.

111. **Berberich, D.W., Hail, M.E., Johnson, J.V., and Yost, R.A.**, Mass-selection of reactant ions for chemical ionization in quadrupole ion trap and triple quadrupole mass spectrometers, *Int. J. Mass Spectrom. Ion Processes*, 94, 115, 1989.

INDEX

A

Acetonitrile reagent, 90
Acetylene, hydrogen abstraction, 99
Acidity, gas-phase, 37–40
Acrylonitrile reagent, 90
Adduct formation, molecular mass
    determination, 3
Alcohols, 125–129
  gas-phase acidities, 40
  hydroxide ion abstraction, 86
Aldehydes, 130–131
Alkadienes, 119–120
Alkanes, 113–115
Alkenes, 115–119
Alkoxide ion abstraction, 136
Alkyl amines, 75
Alkylamines, 137, 139
Alkylation ion reactivities, 186
Alkyl esters, 134–136
Alkyl group displacement, oxide ion
    reaction products, 99
Alkyl halides, halide ion abstraction from,
    86
Alkynes, 119–120
Amines, 137–139
Amino acids and derivatives, 147–150
  Bronsted base reagent systems, 98
  methane chemical ionization mass
    spectra, 81
  methyl chemical ionization MS, 78
Amino acids and derivatives, 147–150
Amino alcohols, stereochemical effects,
    173
Ammonia chemical ionization MS, 73
  alcohols, 125–126
  carbohydrates and derivatives, 151–152
  dicarboxylic acids, 180
  reagent gases, 75
  stereochemical effects, 177
Anions, see Negative ions
Aromatic hydrocarbons, 125
  halogenated, 141–145
  nitrobenzene, 139
  polycyclic, 92, 124
Artifacts, electron capture chemical
    ionization, 94–95
Association reactions
  negative ion, 31–32
  positive ion, 23–24
Associative detachment reactions, 27–28

Atmospheric pressure ionization (API)
    mass spectrometry, 102
  Bronsted acid reagents, 73
  instrumentation, 52–55
Average dipole orientation (ADO) theory,
    10–12, 17, 30

B

Basicity, gas-phase, 32–37
Benzoin methyl ether oxime, 76–77
Biogenic amines, 139
Bromobenzene, 142
Bronsted acid chemical ionization MS
  alcohols, 125
  amino acid and derivatives, 147–149
  carboxylic acids and esters, 133–134
  esters, 134–135
  glucuronides, 154
  halogenated compounds, 141, 145
  nitro compounds, 139
  peptides, 150
  stereochemical effects, 178–179, 183
    chirality effects, 184
    cyclic diols, 176
  stereochemical effects in, 173–174
  tandem mass spectrometry, protonation
    sites, 185
Bronsted acid reagents, 71–82
  carbohydrates and derivatives, 150
  methane, 71–73, see also Methane
    chemical ionization MS
Bronsted acids
  gas-phase acidities, 39–40
  isotope exchange reactions, 169
  proton transfer reactions, 17
Bronsted bases
  isotope exchange reactions, 169–171
  oxide ion reaction products, 98
  reagent systems, 95–98
Butane reagent gas, 72
Butyl acetate chemical ionization, 81

C

Capture collision, 9–10
Capture collision rate constants, 10–11
Capture cross section, 10
Capture rate constant, 10
Carbohydrates and derivatives, 150–156
Carbonyl compound gas-phase acidities, 40

D

E